System Dynamics for Performance Management & Governance

Volume 8

Riccardo Santoni

The Dynamics of Materiality

The Role of Sustainability Accounting in Supporting Decision Making

CED4 System Dynamics Group

Springer

Riccardo Santoni
Business and Law Studies
University of Siena
Siena, Italy

ISSN 2367-0940 ISSN 2367-0959 (electronic)
System Dynamics for Performance Management & Governance
ISBN 978-3-032-19279-0 ISBN 978-3-032-19280-6 (eBook)
https://doi.org/10.1007/978-3-032-19280-6

This Springer imprint is published by the registered company Springer Nature Switzerland AG
The registered company address is: Gewerbestrasse 11, 6330 Cham, Switzerland

Preface

Dear reader,

It is with great joy that I write these forewords, hoping to be able to convey the passion and curiosity that have guided every step of the research behind these pages.

This book tells the story of my Ph.D. journey, including the decision to move with my family and my little daughter to Bergen, to learn the System Dynamics method and apply it to my ideas. From the very beginning, I saw sustainability reporting as something too revolutionary to fit within existing paradigms and methods. There were too many interconnections, too many actors, too many overlooked effects and countereffects.

I felt the need to go beyond the dominant static approach that risks turning sustainability reporting into a mere compliance exercise. I wanted to play my part, to translate a systemic vision into something readable and applicable to business contexts, to help managers dominate the complexity of sustainability performance management.

The question that guided me was simple: *how can sustainability reporting be transformed from a compliance obligation into a strategic lever for change?*

I traced the origins of sustainability and sustainability accounting, explored the history, the milestones, and the academic debates, searching for the point where the right path was lost. Chapter 1 retraces this journey and leads to today's most pressing issue: materiality. What to disclose, how, and for whom? Confusion reigns, companies adopt diverse solutions without knowing whether they are right or wrong. So I asked myself: *is materiality what really matters?*

In Chapter 2, I explored the causal nodes in which materiality is involved. It touches every dimension of business and even goes beyond the company's perimeter. Materiality requires stakeholder involvement and influences how investors allocate resources. Its assessment cannot be confined to the disclosure moment; it needs to be internalised and integrated into decision-making.

That was the turning point of my research, and the moment in which I decided to move to Bergen to attend the System Dynamics course. I learned how to translate narratives into maps, and maps into simulation models. I shared my idea: to use sustainability reporting as a management tool and to give sustainability performance the same polarity as financial performance, convinced that sustainability will be the main source of competitive advantage in the future. And then came the question from my supervisor: *Well, that's fine, but what's the problem?*

The problem was clear: nobody trusts sustainability disclosure anymore. We cannot distinguish what is truly sustainable from what is merely sold as sustainable. Greenwashing is like a green smoke that makes everything opaque.

This challenge shaped Chapter 3: if greenwashing and scepticism are the problems, we have to find their roots, include them in the model, understand their causes, and explain them to the world. That is where the modelling process began. All the pieces of the puzzle were there; they just needed to be placed in the right spot. The model is simple, deliberately generic, but capable of capturing essential dynamics. Because the value lies not only in the simulation results, but also in the modelling process itself and the questions it opens.

Chapter 4 brings these ideas to life through dynamic simulation. It shows how short-term tactics, such as greenwashing, may offer temporary legitimacy but ultimately trap companies in a cycle of distrust and stagnation. It identifies leverage points that can break this cycle: proactive resource allocation, transparency, and the integration of materiality into decision-making. These levers demonstrate that sustainability reporting can become a strategic tool, guiding firms towards long-term value creation.

However, this book is not just about models and simulation; it is about vision. It invites managers and scholars to embrace complexity and rethink sustainability reporting as part of a broader managerial system. In doing so, it offers a call to design the future, not merely to report the past.

Siena, Italy Riccardo Santoni

Contents

List of Figures

List of Tables

Chapter 1
Sustainability Reporting History and Current Challenges

Contents

1.1 The Concept of Sustainability from a Global and Business Administration Perspective

1.1.1 The Path Towards Sustainable Development Goals (SDGs)

On 25 September 2015, 193 member states of the United Nations (UN) agreed on the adoption of Agenda 2030, a shared global plan of action based on 17 Sustainable Development Goals (SDGs) to be achieved by 2030 (United Nations, 2015b). People, planet, peace, and cooperation represent the building blocks of the

R. Santoni, *The Dynamics of Materiality*, System Dynamics for Performance Management & Governance 8, https://doi.org/10.1007/978-3-032-19280-6_1

agreement, often described as a final call to tackle climate change and keep the average global temperature below +2 °C compared to pre-industrial levels. Although progress remains insufficient (United Nations, 2023b), and current geopolitical tensions may fuel further pessimism, the ecological transition has begun, bringing managerial challenges at both institutional and organisational levels. These challenges require active participation from the academic community, which is called to contribute to a sustainable transition (Goolsby et al., 2023).

Agenda 2030 represents the latest outcome of a long-standing global debate on sustainability, which intensified in the post-war period (Klarin, 2018). This section retraces some of the most representative stages of this debate, highlighting key definitions, institutional initiatives, wake-up calls, and appeals for a shift in the global economic paradigm, which have mostly remained unanswered.

A starting point of this historical reconstruction is the Club of Rome experience, one of the pioneering movements in the field of sustainability whose members are intellectuals from various sectors (e.g., scientists, economists, and citizens). Founded in 1968, the Club of Rome brings together individuals committed to shaping a better future for humanity, following the inspiring ideas of Aurelio Peccei, an Italian industrialist, and Alexander King, the Scottish Head of Science at the Organization for Economic Co-operation and Development (OECD).[1] Since the beginning, the Club of Rome aimed at promoting a global and long-term perspective to face intertwined economic, environmental, political, and social global problems. A turning point in the organisation's life was the encounter in 1970 with Jay Forrester, professor at Massachusetts Institute of Technology (MIT) and inventor of the System Dynamics method (Forrester, 1961).

System Dynamics met the needs of the Club of Rome by offering the possibility to use computer models to simulate the mutual effects of interconnected factors acting as variables of a complex system. A team of researchers at MIT applied this method to model exponential growth through five primary factors: population, agricultural production, non-renewable resource depletion, industrial output, and pollution (Forrester, 1971). The results were published in 1972 in the report "The Limits to Growth" (Meadows et al., 1972). The report drew public attention as it condemned the unconstrained growth paradigm, calling the global society to embrace a different life mindset, respectful of the capacity of nature to regenerate resources and absorb pollutants.

The scope of the research was to spread awareness throughout the population that change depended on human commitment and intention. Differently from researchers' expectations, prominent economists gave life to a harsh debate, criticising the System Dynamics scientific approach and contesting the reliability of the forecasts on viability risks for the human species (Nordhaus, 1973). After years of public debates and attempts to explain that the contribution of System Dynamics lay in the explanation of the system structure and the understanding of problems' causes

[1] The information contained in this paragraph is obtained from the website of the Club of Rome organisation: https://www.clubofrome.org/ accessed on 01/09/2025.

and not in forecasting accuracy (Forrester et al., 1974), the research team abandoned the project. Periodic updates have repeatedly recalibrated the model with the latest real data (Meadows et al., 1992, 2004), always underlining the quality of the initial forecasts. More than 50 years later, "The Limits to Growth" survives as an unheard manifesto of warning, with humanity that is experiencing predicted concerns related to pollution and climate change (United Nations, 2023c).

The Club of Rome experience contains a lot of recurrent aspects and problems linked to the tackling of sustainability issues, like the existence of serious warning signs rooted in scientific research findings as opposed to the efforts to undermine their legitimacy carried out by the defenders of the *status quo* situation and business-as-usual behaviour (Reed et al., 2021).

In the same year as the Limits to Growth publication (1972), the "UN Conference on the Human Environment" took place in Stockholm. It was the first world conference focused on the environmental aspects that laid the foundation for the concept of sustainable development. The Stockholm Declaration marked the start of a dialogue between industrialised and developing countries, highlighting the link between economic growth, environmental issues and societal well-being (United Nations, 1972). The creation of the UN Environment Programme (UNEP), the leading global authority on the environment, is the primary outcome of the Conference and shows the leading role that the UN has historically aimed to play in the sustainability context.

In the following years, the UN continued its efforts with the creation of new initiatives such as the foundation, in 1983, of the World Commission on Environment and Development (WCED), aimed at addressing the challenges which remained unanswered after the Stockholm Conference. The WCED ended its mandate in 1987 with the release of the report "Our Common Future" (known as the "Brundtland Report"), which contained the first and most influential definition of sustainable development: the concept of fulfilling present needs without compromising the ability of future generations to do the same.

Despite the growing institutional efforts and the seeming commitment of the member states, history testifies that the majority of sustainability-related promises have constantly been neglected. This also happened to the outcomes of the UN Conference of Environment and Development—known as the "Earth Summit"—held in Rio de Janeiro in 1992. At the Earth Summit, the UN Framework Convention on Climate Change was established, and climate change was recognised as a question of international law. Today, the states that signed the Convention are members of the Conference of Parties (COP), which meets annually to set new goals and measure achievements. One of the most important conferences, the COP3 held in Kyoto in 1997, led to the approval of the Kyoto Protocol, an international treaty to reduce average Greenhouse Gas (GHG) emissions for the period 2008–2012 by 5% compared to 1990 levels. Once again, commitments were not met, with an increase in global scale emissions for the year 2019 of 42% compared to 1990 levels (IPCC, 2023).

The UN Rio conference was also characterised by a significant participation of non-governmental organisations (NGOs) and succeeded in establishing the global

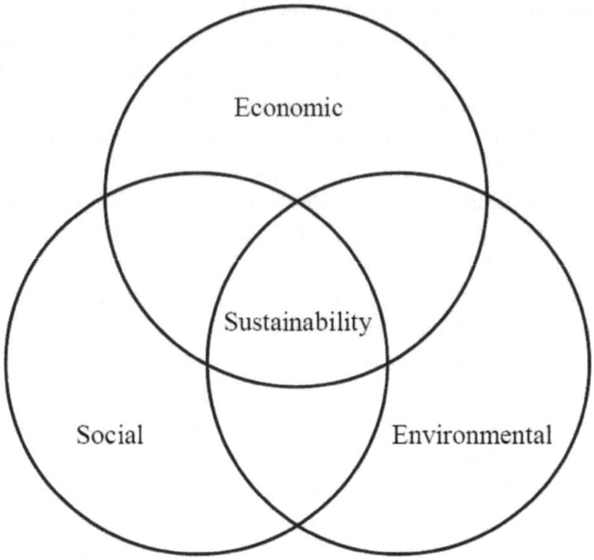

Fig. 1.1 The Triple Bottom Line framework. Source: Author's elaboration of the Triple Bottom Line by Elkington (1994)

extent of environmental and social problems. The economic, social, and environmental dimensions were presented as three critical and interconnected dimensions of the sustainable development challenge, which required rethinking all the production, consumption, life, work, and decision-making processes according to an integrated approach able to balance the weights of the three dimensions (United Nations, 1992).

This tripartite perspective constituted the basis for the well-established framework of the "Triple Bottom Line", proposed by Elkington (1994). According to this framework, activities can be defined as sustainable only when contemporarily addressing the economic, social, and environmental dimensions of sustainability. The graphical display of such a framework is shown in Fig. 1.1 and collocates sustainability at the intersection of the three dimensions.

The Triple Bottom Line framework can be read through the lens of the "strong approach to sustainability", for which no depletion of any sustainability dimension—in particular, the environmental and social ones—should be allowed to reach an advancement in the others—in particular, the economic one (Ball & Milne, 2005). This interpretation is expressed by the ecological approach to economics, which emphasises the importance of maintaining the stocks of natural assets and the impossibility of substituting their functions and services through human-made capital (Costanza et al., 1997; Daly & Farley, 2004). Nevertheless, a strong approach to sustainability exists as opposed to the concept of "weak sustainability". Here, the three pillars are read according to the substitutability principle, with the economic growth and technological development that may increase the overall value of the resources and mitigate the effects of natural resource depletion (Neumayer, 2003).

The latter interpretation has dominated the economic paradigm over the years, postponing the environmental and social needs to the economic ones, trusting in the remedial powers of technological progress (Boos, 2015; Victor et al., 1998). This interpretation distorted the original intent of the Triple Bottom Line, prompting the author to propose a revised framework into a triple helix aimed at regenerating economies, societies, and the biosphere through a new model of value creation (Elkington, 2018).

The importance of adopting a strong and comprehensive approach to sustainability is also underlined by the Planetary Boundaries framework (Rockström et al., 2009). It identifies the limits related to nine environmental areas below which the ecosystems may not be able to self-regulate and may lead to non-linear change on planetary scales. In this context, the consequences of environmental issues can be disruptive and not limited to natural adverse events. Climatic stress directly impacts people's lives, generating immigration and international tensions that we are already starting to experience. For this, the idea of considering sustainability as a nested system of non-substitutable components is essential since environment depletion would mean worse societal conditions, and worse societal conditions mean no development in the economy (Wackernagel & Rees, 1996). Accounting for each system component is necessary to prevent substitution and depletion phenomena.

The beginning of the twenty-first century was an important period for the advancements in the sustainability accounting field, characterised by important public initiatives for measuring and setting sustainability goals (Hardi & Zdan, 2001).

The UN proposed a major initiative with the UN Millennium Development Goals (MDGs) (United Nations, 2000). The MDGs set eight goals, articulated in 21 targets, to be reached in the 2000–2015 period. Six objectives on a total of eight regarded the social sphere, that is, eradicate extreme poverty and hunger, achieve universal primary education, promote gender equality and empower women, reduce child mortality rates, improve maternal health, and combat HIV/AIDS, malaria, and other diseases. The remaining two were about environmental sustainability—articulated in sustainable development principles diffusion, protection of biodiversity, access to safe drinking water, and improvement of slum dwellers' lives—and the development of a global partnership for the goals. Despite the specific orientation of the MDGs towards the reduction of inequalities between the most and least industrialised countries, uneven results were achieved, with the less developed countries not experiencing significant improvement in social conditions (Stuart & Woodroffe, 2016; Vandemoortele, 2011). Also, the environmental goals registered low commitment to the public political agendas (Donat Castelló et al., 2010), and the formulation of the targets revealed to be inadequate to promote cooperation and adapt the sustainability measures to a specific country's needs (Fehling et al., 2013).

In 2012, during the UN Rio de Janeiro conference, the advancement in MDGs was considered inadequate, and the assembly decided to undertake the creation of new goals and targets. Such a challenge led to the issuance, in 2015, of the Sustainable Development Goals (SDGs), first presented at the New York UN Summit. The Summit was followed by the Paris Agreement (COP21), which is considered a historical milestone signed by the agreement of 196 countries to keep the increase in global average temperature below +1.5 °C compared to pre-industrial levels (United Nations, 2015a).

The 17 SDGs comprehend the three dimensions of the Triple Bottom Line framework. The following categorisation attributes each goal to its most representative category:

- Economic dimension: (SDG n. 8) Decent work and economic growth, (9) industry innovation and infrastructure, and (12) responsible consumption and production.
- Social dimension: (1) No poverty, (2) zero hunger, (3) good wealth and well-being, (4) quality education, (5) gender equality, (6) clean water and sanitation, (7) affordable and clean energy, (10) reduced inequalities, (11) sustainable cities and communities, and (16) peace, justice, and strong institutions.
- Environmental dimension: (13) Climate action, (14) life below water, (15) life on land.

In addition, SDG n. 17—partnerships for the goals—focuses on the enhancement of cooperation among the actors (e.g., institutions, governments, NGOs, companies, and citizens) to promote sustainable development.

The subdivision into three categories is not rigid since some goals are multidimensional by nature. SDG n. 6—clean water and sanitation—is one of the most straightforward examples of this situation; drinking water is such an essential good that it is impossible to think of environmental, social, or economic development without access to it. The described interrelationships are still more evident when moving to the sub-articulations of the 17 SDGs: To operationalise them and keep track of the progress, the UN has proposed a system of 169 indivisible targets measurable through 232 indicators (United Nations, 2023a).

Although the SDGs are presented as individually achievable measures, they are part of a network of relationships and mutual trade-offs that discourage a partial analysis, calling governments to a generalised and widespread commitment. Like the previous example of water, the growing negative impacts of climate change require urgent interventions to prevent irremediable consequences also on other social and economic objectives (Adams et al., 2020). Moreover, the SDGs are involved in mutually beneficial relationships, represented by positive externalities, where progress in one field supports advancements in others (Nilsson et al., 2018). This nested nature offers managerial opportunities to improve overall SDG performance by leveraging these synergies. The Millennium Institute Project is a successful experience in this field: Through System Dynamics applications, the project enabled governments and organisations to design strategies and activities to foster interaction among the SDGs, promoting a faster and measurable transition towards sustainability (Pedercini et al., 2019).[2]

The SDGs and the Paris Agreement sound like the last call to limit climate change advancement. The synergies among the SDGs are an excellent opportunity

---

[2] Further information on the Millennium Institute initiatives is available on the organisation's website:

https://www.millennium-institute.org/ accessed on 01/09/2025.

to reveal the mutual benefits that are expected if the diverse actors agree to cooperate to reach the goals, as requested by SDG n. 17.

To prevent the SDGs challenges from becoming another story of failure, as happened to the other valuable efforts portrayed in this first section, a strong involvement of the private sector cannot be postponed anymore (Sachs, 2012). The companies play a complementary role to the government and have the power to promote the desired paradigm shift that institutions alone have not been able to reach in more than 50 years since the Stockholm Conference (Rashed & Shah, 2021). Driving the change in the world of businesses implies intense managerial challenges that impact several aspects of the business administration discipline.

Addressing this call, the following section presents an introduction to the historical stages of the concept of corporate responsibility and the environmental, social, and governance (ESG) dimensions of business administration.

1.1.2 Systemic Perspectives on Corporate Sustainability and Financial Disclosure

The institutional milestones that led to the global agreement on Agenda 2030 reflect a broader societal shift: growing concerns about environmental and social issues have increasingly challenged the promises of well-being tied to economic growth (Biermann et al., 2017).

In response, initiatives such as the UN Global Compact (UNGC), launched in 2000, aimed to align corporate behaviour with societal needs (Williams, 2018). Today, with over 25,000 participating companies, the UNGC stands as the world's largest corporate sustainability initiative. The scope of the project is formulated as a call to companies to align strategies and operations to ten principles concerning human rights, labour, environment, and anti-corruption.[3]

To take actions that advance societal goals, the UNGC initiative provides managerial tools to include the SDGs in business lives and strengthen ESG performance. Concerning this last acronym, ESG is today the most diffused term in the financial markets to indicate the sustainability performance and behaviour of organisations. The addition of the concept of governance to the environmental and social dimensions of the Triple Bottom Line entails considering the combination of actions and controls put in practice to drive the sustainability transition, involving the more managerial dimension of the sustainability challenges (Busco et al., 2005).

Several frameworks have been developed to address the wide range of factors surrounding the concept of sustainability (Elkington, 2018). Some focus on capturing the value that corporate activities generate for the environment and society,

[3] Further information on the UNGC is available on the website of the initiative: https://unglobalcompact.org/what-is-gc accessed on 01/09/2025.

moving beyond purely financial analysis. This is the case of impact-related measures, which account for the broad positive and negative repercussions of business activities and investments on the environment and all stakeholders (Bugg-Levine & Emerson, 2011). Others aim to improve the efficiency of production processes and economic systems, promoting less carbon-intensive activities (Lovins et al., 2005) or the valorisation of waste through circular economy approaches (Murray et al., 2017).

Before the widespread adoption of ESG performance measurement, scoring, and ranking systems, corporate efforts to improve societal well-being were primarily studied within the corporate social responsibility (CSR) research stream (Du et al., 2010, p. 8; Kotler & Lee, 2004). CSR links decision-making to ethical values and respect for employees, communities, and the natural environment, supporting the company's accreditation and legitimisation processes within society (Deegan, 2002; Guthrie & Parker, 1989). Conceptually, CSR represents the most managerial dimension of business ethics (Carroll et al., 2017; Pizzi et al., 2020, p. 20). The broader notion of business ethics helps clarify the deep-rooted differences between business administration and socio-economic disciplines: while the former applies microeconomic principles to the social dimension of business life—enabling the evaluation of complex phenomena and relational dynamics—the latter tends to adopt a more macroeconomic and financial perspective (Cassandro, 1982).

Academic tradition has long recognised the close connections between companies and the social and environmental contexts in which they operate. One of the earliest contributions came from Gino Zappa, who described the firm as a system of coordinated economic activities in which human involvement and resources are essential, with the ultimate scope of meeting societal needs (Zappa, 1927). This human-centric view, later reinforced in his subsequent work (Zappa, 1956), marked a shift away from earlier contractarian perspectives, which saw the firm as a union of capital and administrative activity (Catturi, 2010, p. 58).

Subsequent contributions developed a systemic approach, emphasising the relational and dynamic nature of business organisations (Amaduzzi, 1949; Onida, 1953). In this view, firms are embedded within their broader context, continuously interacting with society and the environment to seize opportunities and manage constraints (Ferrero, 1987; Pizzi, 2022).

A complementary perspective, known as the organicistic theory, conceptualises the firm as a living organism composed of interrelated subsystems. Alberto Ceccherelli described the company as a microorganism within a larger social system, with internal structures functioning like organs (Ceccherelli, 1962, 1967). This view was further developed by scholars such as Riparbelli, Giannessi, Ponzanelli, and Catturi, who defined the firm as a socio-economic entity with specialised functions and an inherent openness to its external environment (Catturi, 2021; Ponzanelli, 1977).

These systemic and organicistic perspectives highlight the firm's continuous search for equilibrium between internal operations and external pressures. They also emphasise the social relationships within the organisation, viewing the workforce not merely as a resource but as a constitutive element of the firm itself. This differs from the profit-maximisation logic typical of capitalist systems and supports a broader concept of value that includes social and socio-economic goals (Di Cagno et al., 2003). This broader view is encapsulated in the notion of "global added value" (Catturi, 2021, p. 575).

The conceptual overlap between these perspectives and the principles of CSR has led some scholars to argue that these early contributions anticipated key elements of CSR theory (Catturi, 2010; Coronella et al., 2018, p. 645). The idea of global added value also challenges the shareholder-centric view, which holds that firms exist primarily to generate financial returns for capital owners (Friedman, 1962; Jensen & Meckling, 1976). Instead, the firm is seen as an open system whose boundaries extend into the social and environmental spheres, requiring alignment between corporate behaviour and stakeholder expectations (Catturi, 2010, p. 71).

This view aligns closely with stakeholder theory, which posits that successful firms must create value for a wide range of actors—including customers, suppliers, employees, communities, financiers, and shareholders—and that managers must balance these interests to achieve organisational legitimacy (Freeman, 1984).

Both shareholder and stakeholder theories can be read as specific types of the broader agency theory, for which the company is a contract between two parties, the principal and the agent. The former delegates the latter to act on its behalf to pursue its interests. Depending on the theoretical approach selected, managers (i.e., agents) have to account for their actions to shareholders or the plurality of stakeholders (i.e., principals) (Freeman, 2008). In either case, problems related to opportunistic behaviours may emerge, with agents that may take advantage of information asymmetries due to their privileged position inside the business (Jensen & Meckling, 1976).

Over time, regulators, researchers, and practitioners have continuously directed their efforts to reduce the difficulties related to information asymmetry and ensure the reliability of business financial communications, giving life to a set of national and international reporting rules that constitute the subject of financial accounting (Alexander et al., 2017; Dennis, 2014). The spread of common standards and procedures provides capital markets with reliable, high-quality, and comparable information that supports everyday investment decisions (Di Pietra, 2015).

In financial statements, environmental and social factors represent causes of estimation uncertainty in the values of assets and must be evaluated if they have the potential to undermine business continuity (International Accounting Standards, IAS 1, par 25). In particular, climate risk may jeopardise future cash flows, impacting the process of fair value estimation of assets and stocks (IAS 36), or may cause the adoption of fiscal, normative, or financial incentives—or disincentives—impacting some business activities (e.g., excess profit tax, carbon tax, green loan) (Strada, 2023). Differently, the other environmental and social aspects not directly affecting fair value estimations and business continuity might not be displayed inside the financial statement but, if relevant, should be described in the management report (Adamo & Di Cagno, 2020).

What emerges is that the importance that financial statements attribute to social and environmental factors is limited to the financially quantifiable direct—usually negative—impacts that such factors have on the capacity of a firm to continue its business-as-usual proposition of value. Such a limited approach to disclosure is unable to capture the perpetual search for balance between the company and the external context (Catturi, 2021), falling short of attributing legitimation to business activities and obtaining stakeholders' recognition (Buhmann, 2016).

It is important to note how the selection of adequate indicators and targets is essential to setting proper goals to orient business activities through managerial control systems (Merchant, 1982). In this context, the financial metrics alone may be ineffective in contributing to the quantification of environmental and social impacts, preventing the possibility of using proper measures to guide the management towards a broader concept of value. This may generate pathological consequences: given the marginal contribution of the ESG factors to financial results and the business orientation to the maximisation of profits stimulated by financial indicators, ESG factors will be ranked at the bottom of the business hierarchy and have low weight in the decision-making process. This risky perspective moves businesses away from their natural goal, which should be aligned with the societal one of achieving the common good (Coda, 2012).

Nevertheless, the need to develop financial measures to account for ESG performance clashes with operational and theoretical challenges linked to the multifaceted nature of the sustainability factors. Indeed, economic goods are susceptible to monetary evaluation, a fundamental aspect for homogenising the accounting system (Hummels & Timmer, 2004). On the contrary, measuring a company's ESG performance requires taking into account many aspects belonging to diverse scientific fields (e.g., measures for GHG emissions, water consumption, and employee satisfaction) for which it is often hard to determine the monetary value (Adams & McNicholas, 2007). The set of additional competencies that sustainability accounting scholars, preparers, and users should possess stimulated academic research to define sustainability accounting as a young and independent discipline (Schaltegger et al., 2013).

Building on the previous sections, which explored the institutional evolution of sustainability and its integration into managerial thinking, the next part of the chapter traces the development of sustainability accounting as a distinct discipline—from its early foundations in the late twentieth century to its current role in shaping corporate reporting practices.

1.2 The Historical Stages of Sustainability Reporting

1.2.1 Social and Environmental Disclosures in the Twentieth Century

Historical research in the field of social and environmental accounting remains relatively limited (Pizzi, 2022, p. 42). Early conceptualisations of this discipline (see Gray et al., 1987) framed it as a process through which organisations communicate the broader social and environmental consequences of their economic activities, moving beyond the traditional focus on financial reporting to shareholders. This perspective is grounded in the assumption that companies have responsibilities that extend beyond profit generation.

Such views emerged after a period of profound social changes, as it was in the 1970s, characterised by the rise of trade union strikes and demands for improved working conditions. As discussed in Sect. 1.1.1, this same decade also witnessed the rise of environmental awareness, exemplified by the publication of The Limits to Growth in 1972. However, environmental concerns did not trigger the same level of mass mobilisation as social issues.

The social dimension has long been at the forefront of the international debate on sustainability reporting. In the 1970s, initial isolated attempts to account for social externalities produced by corporate practices began to surface. These early examples of social reporting stimulated academic debate around the nature and purpose of this emerging practice. In some cases, social reports were interpreted as extensions of financial statements, developed through alternative evaluation processes (Cassandro, 1989); in others, they were seen as tools to make corporate behaviour more comprehensible to external stakeholders (Vermiglio, 1984). A more proactive interpretation also emerged, envisioning social reports as instruments to foster interaction and engagement between companies and their stakeholders (Salvioni, 1992).

Gray et al. (1988) retraced this debate, observing that the first 15 years of CSR disclosure practices were marked by a lack of conceptual clarity and theoretical consensus. Rather than being guided by a unified framework, CSR evolved through a fragmented and often contested terrain, reflecting deeper tensions within accounting thought itself. The discipline struggled to define its boundaries, its objectives, and even its legitimacy, especially in contrast to the well-established norms of financial reporting.

In this context, CSR questioned the primacy of investors and financial markets as the sole stakeholders of corporate information, opening up space for broader societal concerns. It implied a recognition that organisations operate within a web of social expectations and responsibilities, calling for a deeply political change that challenged the conventional view of accounting as a passive, objective tool. From this perspective, CSR could be interpreted as a form of social contract through which companies respond to external expectations and seek legitimacy in society.

This shift reflects a broader transformation in the perceived role of accounting itself. Indeed, CSR introduced the idea that accountability encompasses society at large, meaning that the legitimacy of corporate behaviour should be earned through the provision of meaningful information. In this way, accountability becomes embedded in a dynamic social contract that evolves over time, shaped by changing societal values, political pressures, and public expectations.

By the final decade of the twentieth century, these efforts had evolved into a broader wave of experimentation aimed at capturing the full spectrum of externalities generated by business activities (Unerman et al., 2018, p. 504). This period was also marked by growing concern for social and environmental issues and by improvements in the structural conditions for sustainability reporting (Larrinaga &

Bebbington, 2021, pp. 166–167). On the industry side, organisations increasingly recognised the need to disclose social and ecological information to a wider audience beyond shareholders. This evolution was shaped by a convergence of earlier reporting traditions (e.g., employee, social, and environmental reporting) into a reporting structured around the Triple Bottom Line framework (Buhr et al., 2014). This shift reflected a growing recognition that corporate performance required an integration of environmental, social, and economic indicators into a unified perspective, even though the initiatives still remained largely voluntary and unregulated.

In this period, the proliferation of sustainability disclosures, often inconsistent and difficult to compare, highlighted the need for common guidance. Indeed, the lack of uniform metrics and assurance mechanisms undermined the credibility and usefulness of sustainability reports, prompting calls for regulatory involvement. Aiming to enhance transparency, comparability, and stakeholder engagement, knowledge-based experts (i.e., the "epistemic communities" of academics and practitioners) and regulators (e.g., the European Union's Towards Sustainability Programme; Hillary, 1995) laid the groundwork for the development of social and environmental measurement and reporting standards.

These efforts culminated in the establishment of the Global Reporting Initiative (GRI) (Bebbington, 2021), which quickly gained traction among corporations and institutional investors, becoming a cornerstone of sustainability reporting practice.

1.2.2 *The Global Reporting Initiative (GRI)*

In 1997, as a result of the social, institutional, organisational, and academic debate, the GRI was founded.[4] The organisation aimed to create the first accountability mechanism to spread shared environmental conduct principles, then broaden to comprehend even social, economic, and governance issues. To reach this goal, in 2000, the GRI Guidelines G1 were released, providing the first global framework for sustainability reporting.

The idea behind the launch of GRI guidelines was to build consensus on a sustainability reporting framework to create a process of growth, affirmation, and improvement similar to the one that affected the story of financial reporting (Etzion & Ferraro, 2010). As concerning the relationship between these two types of disclosure (i.e., sustainability and financial), GRI aimed at complementing financial reports with a forward-looking orientation of key value drivers such as human capital, management and governance system, environmental risks and liabilities, and innovation capacities (GRI, 2002, p. 68). Complementarity, openness, transparency, and alignment with societal and organisational interests (e.g., with the UNGC project) made the initiative well accepted and led to widespread adoption among all

[4] The information on GRI is available on the organisation's website: https://www.globalreporting.org/about-gri/mission-history/ accessed on 02//09/2025.

types of companies, in particular by the largest global ones (Levy et al., 2010). This enabled, for the first time, the legitimacy and the convergence of sustainability reporting practices (Larrinaga & Bebbington, 2021, p. 173). Other reasons for the fast legitimation of the GRI guidelines can be found in the analogies with financial reporting procedures; to this point, GRI explicitly referred to the financial accounting history and aimed at starting a long-term endeavour to build a globally accepted reporting framework based on sustainability performance indicators (GRI, 2002, p. 1). Together with these efforts to get closer to business, GRI put a central focus on stakeholders and the concept of accountability of corporations' impacts on the environment and society (GRI, 2002, pp. 4–5).

The continuous upgrade of always more punctual and pervasive guidelines is a critical factor that helped GRI maintain its dominant position (de Villiers et al., 2022) in the sustainability reporting field. This continuous evolution is well illustrated by the timeline of GRI's development. Since its foundation in 1997, GRI has progressively expanded its scope and influence. The first version of the guidelines (G1) was launched in 2000, followed by successive updates (G2 in 2002, G3 in 2006, and G4 in 2013), each introducing more refined principles and metrics such as the guidance for how to prioritise issues and define reporting boundaries (i.e., materiality assessment) and the introduction of tests to ensure that each issue was correctly defined and tackled (GRI, 2013).

In 2016, GRI transitioned from guidelines to formal standards, marking a significant shift in structure and ambition. The listing model typical of the guidelines was substituted by a new modular approach, in which universal and common standards (i.e., requirements, organisation description, guidance on materiality assessment) were divided from topic-specific ones, leaving the possibility for each adopter to focus on the identified relevant topics. This modular system was further enhanced in 2021 with the publication of revised Universal Standards and the launch of the first Sector Standard, dedicated to the oil and gas industry. Alongside these technical developments, GRI also built institutional capacity through initiatives such as the Certified Training Partner Program (2008), the GRI Academy (2020), and collaborations with the UN Global Compact to align reporting with the Sustainable Development Goals (SDGs), particularly SDG n. 12.6 on corporate transparency.

Despite these significant advancements in structure, scope, and institutional support, the actual uptake and transformative impact of GRI standards within corporate practices remains uneven. The investigation carried out by KPMG (2017, p. 28) confirmed GRI's dominant position, with 75% of the reports published by the world's 250 biggest firms adopting the standard. Also, the majority of academics agree with the benefits of GRI adoption in terms of comparability of reporting structure, information availability, stakeholders' involvement, and relevance of information derived by the materiality assessment (Adams et al., 2022; Chen et al., 2015).

However, a considerable number of companies persist in maintaining business-as-usual practices (Jollands et al., 2018), highlighting a gap between the aspirations of sustainability frameworks like GRI and the practical implementation within corporate strategies and operations (Wiseman, 1982). One possible cause may be found in the mimetic reasons (Dimaggio & Powell, 1983) behind the choice of adoption of sustainability reporting: In voluntary adoption cases, the organisations often imitate the reporting practices of other firms, adjusting only the disclosure instead of changing the production process (Carungu et al., 2020). This fosters the lobbying power of big companies that succeed in shaping sustainability reporting according to their interests, often oriented to defending their business-as-usual approach and achieving legitimisation from stakeholders (Calabrese et al., 2022; García-Meca & Martinez-Ferrero, 2021; Levy et al., 2010).

This compliance-oriented (Borgert et al., 2018) and legitimation-driven (Calabrese et al., 2022) approach to sustainability accounting, often motivated by companies' desire to preserve market dominance (Cho et al., 2015; Milne & Gray, 2013), has contributed to the limited effectiveness of sustainability accounting in supporting a genuine transition towards sustainability. Critical research streams emphasise that sustainability is a global and systemic issue, deeply rooted in the economic growth paradigm that underpins corporate behaviour. From this perspective, it becomes problematic to consider companies as capable of truly accounting for sustainability. Indeed, sustainability accounting has been described as a "narrative of un-sustainability" (Gray, 2010), highlighting its role in reproducing rather than challenging unsustainable practices.

This view supports the idea that the limitations of sustainability standards and guidelines are not the primary cause of the delay in transitioning to more sustainable production and consumption models. Instead, the issue lies in the structural constraints of corporate logic and the instrumental use of reporting for reputational purposes. Nevertheless, recent years have seen the emergence of new standard setters offering goal- and issue-oriented tools, as well as reporting frameworks tailored to the needs of specific stakeholder groups. These initiatives aim to enhance the transformative potential of sustainability reporting by embedding it more deeply into managerial decision-making. The following subsection presents an example of such efforts, represented by the International Integrated Reporting Council (IIRC) framework, which seeks to integrate sustainability information into the core business model of reporting organisations.

1.2.3 The International Integrated Reporting Council (IIRC) Framework

The development of the International Integrated Reporting Council (IIRC) framework, officially released in 2013 (IIRC, 2013), began in 2010 as a collaborative effort involving key actors in the sustainability reporting landscape, including the GRI. The initiative emerged from the growing need to consolidate financial and non-financial disclosures into a single, coherent document capable of communicating how organisations create value over time (de Villiers et al., 2014, p. 1046). The IIRC proposed an Integrated Report as the organisation's primary disclosure tool, encompassing the most material economic, social, and environmental outcomes, to gradually replace the idea of producing two separate financial and sustainability reports.

While the IIRC shares several principles with the GRI, such as stakeholder engagement, materiality, and comparability, it diverges in its intended audience and strategic orientation. The Integrated Report is primarily designed to inform providers of financial capital, focusing on how an organisation's activities affect its ability to generate value in the short, medium, and long term. This investor-centric approach has drawn criticism from scholars concerned with ecological integrity and systemic sustainability (e.g., Milne & Gray, 2013).

A distinctive feature of the Integrated Reporting framework lies in its innovative categorisation of sustainability elements. Organisations are required to articulate how they create, preserve, or erode value over time with respect to six interconnected forms of capital: financial (e.g., funds), manufactured (e.g., physical infrastructure), intellectual (e.g., knowledge and systems), human (e.g., skills and well-being), social and relationship (e.g., networks and trust), and natural (e.g., environmental resources). These capitals are conceived as inputs that, through business activities, are transformed into outputs and outcomes. For each capital, the organisation should account for the value generated or lost, assessing performance across short-, medium-, and long-term horizons, and providing connected information on cause-effect relationships between strategic decisions, resource use, and performance outcomes (IIRC, 2021). This logic focuses on a more holistic representation of value creation, aimed at embedding sustainability into core business processes to overcome the limitations linked to compliance-oriented disclosures.

This systemic view implies that the capitals are not isolated silos but part of a closed and dynamic cycle, where trade-offs and synergies among them must be considered. The framework encourages organisations to reflect on how strategic decisions, resource allocations, and risk assessments affect the balance and sustainability of these capitals (Santoni, 2023).

Central to this approach is the concept of integrated thinking, which involves a deliberate assessment of the relationships among an organisation's units and resources to support cohesive decision-making (IIRC, 2021, p. 3). This concept closely resembles the definition of a management system as a unified structure of processes and tools used to formulate strategy, implement actions, and monitor effectiveness (Kaplan & Norton, 2008, p. 1).

Promoting integrated thinking means adopting a holistic perspective on the business model, embedding targets and performance indicators for each capital into accounting and control systems (Busco et al., 2021; Malafronte & Pereira, 2021). This integration enhances employees' sense of belonging (i.e., human capital), improves control effectiveness, and fosters internal change processes (Busco et al., 2021; Dimes & de Villiers, 2020). It also strengthens awareness of intangible inputs in value creation (Herath et al., 2021), improves risk management routines (Macias & Farfan-Lievano, 2017), and supports the pursuit of higher and greener performance levels (Di Vaio et al., 2020).

Nonetheless, the Integrated Reporting initiative has not succeeded in transforming the dominant economic paradigm. The potential benefits envisioned by a substantial body of literature (e.g., Adams, 2015; de Villiers et al., 2014) often remained unrealised (Melloni et al., 2016). Beyond initial scepticism (Flower, 2015; Thomson, 2015), several conceptual and operational challenges emerged. Conceptually, the six-capital model introduced a disclosure language that differs from other frameworks, reducing comparability and limiting references to the SDGs (Santoni, 2023, pp. 58–62). Operationally, the vagueness of integrated thinking and the lack of practical guidance made the reporting process appear complex, costly, and demanding in terms of managerial commitment (Dumay et al., 2017; Feng et al., 2017, p. 348; Gibassier et al., 2018).

These factors contributed to a fragmented global adoption, with strong uptake in countries such as Japan and South Africa and limited diffusion in North and South America (Gibassier et al., 2019, pp. 20–21). In 2021, the IIRC merged with the Sustainability Accounting Standards Board (SASB), forming the Value Reporting Foundation (VRF), now consolidated into the International Financial Reporting Standards (IFRS) Foundation. While this consolidation may dilute the identity of Integrated Reporting as a standalone framework, its influence on sustainability accounting and performance management remains significant and enduring.

The following subsection introduces other major sustainability frameworks and explores the drivers behind recent consolidation trends.

1.2.4 Other Sustainability Standard Frameworks

The previous two sections have illustrated how the ambition to make sustainability performance both universally accountable and embedded within business routines led to the emergence of two major innovations in the field of sustainability accounting: the GRI, focused on disclosure practices, and the IIRC, oriented to managerial integration. However, due to the complexity and breadth of sustainability issues, different stakeholders have different expectations and priorities related to business disclosure (AlWaer et al., 2008).

This could be the case of investors, whose desire is represented by easy access to quick and comparable information (e.g., shared indexes and metrics); banks, looking for merit scores linked to a business commitment to sustainability; institutions, hoping for a cleaner economy and verifiable disclosures; or consumers, interested in tangible aspects of sustainability (Stoughton & Ludema, 2012). At the same time, to be welcomed by the adopters, guidelines must offer understandable procedures without being excessively costly.

The sum of all the legitimate interests of the parts may cover the multi-faceted nature of sustainability without neglecting any detail or perspective. On the other hand, a concrete risk of creating uneven and fragmented views—composed of many isolated, narrow perspectives (Gray, 2010)—concretely emerged with the birth of several independent procedures aiming at satisfying the information needs of some specific stakeholders.

This section offers a brief review of some of the most relevant initiatives linked to environmental and social disclosure, highlighting how these frameworks have contributed to offering targeted guidance on specific topics. While not exhaustive, the overview aims to illustrate how such initiatives respond to the diverse expectations of stakeholders and help address the complexity of sustainability by providing structured approaches to distinct issues.

One of the first disclosure systems to be launched was the Carbon Disclosure Project (CDP). Established in 2000, 3 years after the GRI foundation, CDP is a not-for-profit organisation with headquarters on all continents that focuses on the disclosure of environmental information. Initially centred only on carbon and GHG emissions disclosure and reduction, CDP has enriched its guidelines propositions over time by including metrics to help companies improve and account for water and forest usage performance. It aims to widen its guidelines to cover all climate aspects. In addition to this, the organisation explicitly provides recommendations for the accountability of public sector administrations represented by cities and governments. On its website, the organisation counts more than 24,800 adopters among companies and around 1000 among cities.[5]

A different perspective is adopted by another significant standard setter, already mentioned at the end of the previous section. The SASB was founded in the USA in 2011 as a non-profit organisation to link the interests of businesses and investors around corporate sustainability disclosure. To reach this scope, SASB standards provide companies with quantitative and qualitative metrics to disclose the most material challenges for each industry-specific topic. Each metric is accompanied by technical protocols—to ensure the understanding of definitions and the correctness of measurements—and activity metrics—to normalise data and facilitate comparison among companies. All the dimensions of sustainability are covered by the standards and are organised under the categories of environment, social capital, human capital, business model and innovation, and leadership and governance.[6] The proposed disclosure is investor-oriented; organisations should provide information related to the financial repercussions of ESG aspects of business life in terms of risks and opportunities. Such indicators should fulfil the function—made by financial and economic key performance indicators (KPIs) of the financial statements—of orienting investment decisions according to sustainability performance (Busco et al., 2020; Pizzi et al., 2023). In conclusion, SASB Standards are targeted to support organisations in identifying the most material metrics and KPIs related to their activities (i.e., industry), facilitating the measurement and disclosure of their sustainability performance. In 2021, recognising the complementary nature of their

---

[5] Links to the CDP website, accessed on 03/09/2025:

- https://www.cdp.net/en for general information about the organisation;
- https://www.cdp.net/en/about for information on the number of adopters.

[6] Links to the SASB website, accessed on 03/09/2025:

- https://sasb.org/about/ for general information about the organisation;
- https://sasb.org/implementation-primer/ for information about the Standards adoption.

managerial and measurement orientations, the IIRC and SASB merged to form the Value Reporting Foundation (VRF). Despite this union, the two frameworks maintained a certain level of independence. The VRF was later consolidated into the International Sustainability Standards Board (ISSB), marking a further step towards the harmonisation of global sustainability reporting. The implications of this consolidation will be discussed in detail in Sect. 1.3.

Another primary actor in the sustainability reporting debate is[7] the Task force on Climate-related Financial Disclosure (TCFD). In 2015, the Financial Stability Board[8] published a proposal to the G20 for the creation of voluntary disclosures targeted to address the financial risks that climate change brings to companies. After that publication, the TCFD was established to provide recommendations for consistent disclosures that support financial markets in making informed capital allocation decisions by taking climate-related risks into account.[9] The first set of guidelines, launched in 2017, encompasses four areas of disclosure: (1) governance—board evaluation of climate-related risks and opportunities; (2) strategy—long- and medium-term actions to mitigate impacts and improve resilience; (3) risk management—identification of risks and scenario analysis; and (4) metrics and targets adopted. Risks are divided into two major categories: transition risks, related to financial and reputational risks of the adaptation process, and physical risks, driven by climate events or changed climate conditions. In turn, several areas of opportunity are identified in the improvements of resource efficiency, energy use, products and services offered, markets served, and business resilience (TCFD, 2017). Although the focus on investors and the requests for strategic long-term intervention may be aligned with the Integrated Reporting framework, a fundamental difference resides in the role of the capitals (i.e., resources). While Integrated Reporting aims to measure the impact of business activities on capital improvement or depletion over time, TCFD focuses on the concept of dependence on specific resources and how climate change may put at risk their availability (O'Dwyer & Unerman, 2020). With the release and diffusion of the guidelines, the mission of the TCFD was considered completed, the organisation disbanded in 2023, and the IFRS Foundation took over the monitoring of the progress of companies' climate-related disclosures.

Together with the most recognised sustainability reporting frameworks mentioned above, the international panorama is populated with hundreds of

[7] Although the TCFD association disbanded in 2023, the present simple "is" is maintained because of the important role that the TCFD recommendations are still playing in the sustainability reporting scenario.

[8] The Financial Stability Board is an international body that comprehends representations of the highest financial institutions from all the G20 Countries. It promotes international financial stability and coordinates national financial authorities also through the promotion and supervision of standard adoption. Further information is available on the institution's website: https://www.fsb.org/about/ accessed on 03/09/2025.

[9] Further information on TCFD is available on the institution's website: https://www.fsb-tcfd.org/about/ accessed on 03/09/2025.

topic-specific guidelines available to organisations (Gutterman, 2021). In the environmental sector, the Greenhouse Gas Protocol for carbon emissions (GHG Protocol) supplies highly recognised GHG accounting standards and calculation tools adopted by 97% of the 500 world's biggest companies[10]; according to the protocol, organisations should account for carbon emissions at three different levels, called scopes: Scope 1 refers to direct emissions under the control of the company, Scope 2 is about indirect emissions linked to electricity, steam, and power production, Scope 3 is linked to the indirect emissions of the corporate value chain, from the initial purchase to the final disposal of goods.

Among the most important social sector guidelines, the ISO 26000 standard by the International Standard Organisation (ISO) offers guidance on the topic of social responsibility, identifying seven core subjects related to organisational governance, human rights, labour practices, the environment, fair operating practices, consumer issues, and community involvement and development. A higher focus on working practices is proposed in the SA 8000 standards by Social Accountability International (SAI), centred on child-forced or compulsory labour, health and safety, freedom of association, discrimination, disciplinary practices, working hours, remuneration, and management system.[11]

In addition to the spontaneous, market-driven publication of guidelines, the settlement of new standards (or the success of existing ones) is often a consequence of national or supranational initiatives aiming at improving sustainability accounting practices, like in the case of the TCFD settlement.

One of the first examples of this regulatory activity is the South African experience, in which, since 2010, all the Johannesburg Stock Exchange member organisations have been required to publish an "integrated report". This requirement led to the beginning of local consultations, followed by the development of the IIRC guidelines[12] (Sukhari & de Villiers, 2019).

Some years later, the European Union (EU) issued Directive 2014/95/EU on the disclosure of non-financial and diversity information by certain large undertakings and groups, commonly referred to as the Non-Financial Reporting Directive (NFRD). By requiring Member States to transpose the directive into national law, the EU introduced a mandatory approach whereby public interest entities and large organisations are subject to reporting obligations. These organisations must disclose information about their business model, related risks, implemented policies, and other provisions defined at the national level.

For example, Italy implemented the directive through Legislative Decree 254/2016, which requires companies to report on environmental, social, and

[10] Source: GHG Protocol website: https://ghgprotocol.org/ accessed on 03/09/2025.

[11] Further information on GHG, ISO, and SAI protocols is available on the respective institutions' websites: https://ghgprotocol.org/; https://www.iso.org/iso-26000-social-responsibility.html; https://sa-intl.org/; all the websites accessed on 03/09/2025.

[12] The Integrated Reporting Committee of South Africa was set in 2010 to offer guidance to organisations on preparing an integrated report. Further information is available on the committee website: https://integratedreportingsa.org/about/our-history/ accessed on 03/09/2025.

labour-related matters relevant to their business, following the "comply-or-explain" principle (Carungu et al., 2020, p. 452). Notably, the directive does not mandate the use of any specific reporting standard. In practice, most companies rely on the GRI framework—indeed, the entire sample of Italian companies observed in the Deloitte (2022) survey adopted GRI guidelines—but published reports often reflect a combination of multiple standards.

Another relevant case is the process started in 2021 by the Securities and Exchange Commission (SEC), the institutional body in charge of the surveillance of the United States Stock Exchange. The SEC approach identified two elements as the most critical for the sustainability accounting discipline: the volume of GHG emissions and the financial impacts of climate-related risks. In line with this perspective, the SEC decided to explicitly refer to the GHG Protocol for carbon accounting and to the TCFD for addressing climate risks and opportunities. US-listed companies will be required to disclose Scope 1, 2, and 3 emissions and to illustrate the risk management process implemented to counteract climate change (Pizzi, 2022, pp. 82–84).

The increasing specificity and granularity of sustainability frameworks have enabled organisations to address a wide range of environmental, social, and governance issues, offering tailored guidance for different sectors, topics, and stakeholder expectations. This diversity reflects the multi-faceted nature of sustainability and the legitimate need to capture its complexity without ignoring any dimension.

However, such richness also brings challenges. The proliferation of frameworks and metrics risks undermining comparability across organisations and jurisdictions, especially when disclosures are voluntary and context-specific. Moreover, the coexistence of multiple standards can fragment the overall vision of sustainability, generating isolated perspectives and artificial boundaries between issues (Gray, 2010).

These tensions have stimulated academic and institutional debate on the role of mandatory versus voluntary approaches to sustainability reporting. From a business perspective, mandatory reporting has been associated with increased adoption and reduced information asymmetry (Venturelli et al., 2020), but also with higher compliance costs and operational complexity, potentially affecting profitability (Cupertino et al., 2022; De Micco et al., 2020). Furthermore, the integration of financial and non-financial information in a way that enhances comparability often requires standardised procedures, which may limit flexibility and the disclosure of context-sensitive performance (Christensen et al., 2021; Gatti et al., 2019). As a result, some firms may perceive these constraints as a barrier to communicating their sustainability achievements, reducing reporting to a compliance exercise (Cordazzo et al., 2020).

Despite these limitations, institutional commitment remains essential to promote sustainability accountability across markets, enabling capital allocation and consumer choices based on sustainability attributes. Yet, the coexistence of multiple standards—such as Integrated Reporting in South Africa, freedom of choice in Europe, or the GHG Protocol and TCFD in the USA—risks reinforcing fragmentation and hindering global convergence.

After more than 20 years since the foundation of GRI, the sustainability reporting landscape appears to be at a turning point (KPMG, 2020). The abundance of metrics and frameworks offers the potential to measure a wide array of sustainability dimensions, suggesting that sustainability accounting could, in theory, capture sustainability itself. However, the lack of coordination and institutional oversight has led to tensions, including conflicts between standard setters (Fransen & Conzelmann, 2015). Even among companies using the same framework, comparability remains limited due to inconsistent adherence, selective indicator uses, voluntary report opacity, and heterogeneous data quality (Boiral & Henri, 2017; Sitnikov & Bocean, 2017).

In response to these challenges, efforts to build a more coherent and harmonised reporting landscape are underway. The next section explores the emergence of global standardisation initiatives and their implications for the future of sustainability accounting.

1.3 Standardisation of Guidelines and Frontiers of Sustainability Accounting Research

The previous section gave an overview of the guidelines and frameworks that more significantly impacted the sustainability accounting scenario in the last two decades. As discussed, divergencies among frameworks' scopes precluded the possibility of identifying one standard as "winner" or "better than the others", with some guidelines adopting broader perspectives and others keen on their niche of specialisation. Aware of these differences, the standard setters put efforts into offering linkage documents and ensuring that the data provided by an organisation could cover multiple reporting requirements.[13] In theory, these laudable alignment attempts could have led to an ideal situation in which companies shared the commitment to sustainability issues and had the opportunity to use topic-specific guidelines (e.g., risk management, GHG emissions, job security) according to their materiality assessment and industry requirements. In line with this approach, Directive 2014/95/EU aimed to reach similar outcomes by requiring mandatory disclosure for big European firms based on a principle-based approach that did not set any specific obligation on which framework to use (La Torre et al., 2020).

Both the publication of linkage documents and the EU's principle-based approach represent notable attempts to harmonise the sustainability reporting landscape. Harmonisation can be defined as the sum of processes based on flexible regulatory

[13] The GRI website contains a section in which standards linkage documents are published. Examples are represented by GRI-SASB, GRI-TCFD, GRI-CDP alignment procedures. More information can be found at:

https://www.globalreporting.org/how-to-use-the-gri-standards/global-alignment/ accessed on 03/09/2025.

provisions with the scope of making the accounting information homogeneous (Di Pietra, 2003; Pizzi, 2022, pp. 72–73).

The harmonisation attempts proved to be ineffective, with critics of the EU experience linked to the low level of quality and comparability of information, the small number of undertakings subject to mandatory reporting, and the promotion of a mimetic approach that led to a more professional and compliant, but less integrated, sustainability disclosure (Pizzi, 2022, pp. 59–63). In 2018, the EU initiated public consultations to review and update the Directive formulation. Stakeholders called for a new reporting model that would enhance the comparability of information.[14] At the same time, requests for more convergence in the sustainability accounting discipline started to be promoted by global accountancy associations and operators in the economic and financial field (e.g., Accountancy Europe, 2019; World Economic Forum, 2020; and the call from the International Federation of Accountants[15]).

Differently, Adams and Abhayawansa (2022) consider standardised metrics useful to serve investors' needs but not to contribute to sustainability. The authors lead back to political and power scopes—aimed at delegitimising stakeholder-oriented approaches—the sense of urgency around the call for a compelling harmonisation. Although best practices of sustainable strategy integration and positive market value recognition exist (see, e.g., Broadstock et al., 2021 and Santoni, 2023), it is undeniable that most corporate engagement with environmental and social issues tends to be driven more by symbolic actions and strategic intent than by substantive initiatives (van der Waal & Thijssens, 2020). To this point, any initiative capable of promoting the desired paradigm shift would finally be welcomed by academics, preparers, and users.

The requests for a more regulatory commitment to guideline unification go beyond the concept of harmonisation since further intervention would mean higher grades of rigidity, constituting the definition of standardisation (Di Pietra, 2003). Standardisation can be reached through normative provision (i.e., *de jure*) or by the proposition of commonly accepted accounting practices (i.e., *de facto*) (Pizzi, 2022, p. 74; Tay & Parker, 1990). These two approaches characterised the efforts proposed by the European Union and the IFRS Foundation, which have since consolidated into concrete reporting frameworks. The following subsections explore how these initiatives have taken shape and the implications they carry for the future of sustainability reporting.

[14] The directive proposal 2021/0104 (COD) summarises what emerged during stakeholder consultations.

[15] The call for a new sustainability standard board is available at:
https://www.ifac.org/knowledge-gateway/contributing-global-economy/discussion/enhancing-corporate-reporting-way-forward accessed on 03/09/2025.

1.3.1 The Standardisation Process in Europe: A Focus on the Corporate Sustainability Reporting Directive (CSRD)

Driven by a desire to become a global leader in sustainability, the EU is at the forefront of publishing policies and initiatives to support the green transition (Sommer & Hain, 2017). The previously mentioned NFRD 2014/95/EU, which made sustainability reporting mandatory for large European companies, exemplifies this commitment. Another fundamental initiative is the European Green Deal, a comprehensive set of policy initiatives aiming to achieve a "just transition" of the European economy by 2050. To reach climate neutrality by 2050, the Green Deal focuses on three initiatives: "financing, enabling, and supporting" sustainable investments and projects (European Commission, 2020).

A fundamental step to reach these aims is ensuring corporate accountability to investors, civil society organisations, consumers, and other stakeholders (European Commission, 2024). In this context, Directive 2022/2464, known as the Corporate Sustainability Reporting Directive (CSRD, European Union, 2022), has developed standardised, comparable, reliable, and relevant sustainability information, placing sustainability reporting on an equal footing with financial reporting, in line with the scope of the directive proposal published on 21 April 2021.[16]

The preparatory phase to reach this *de jure* standardisation (Di Pietra, 2003) was commissioned to the European Financial Advisory Group (EFRAG), which engaged a broad range of stakeholders, including banks, financial institutions, consultants and auditors, ESG rating agencies, NGOs, academics, and the companies affected by the normative. With reference to the latter, the CSRD is progressively extending its scope to an increasing number of undertakings (Pizzi, 2022). As of fiscal year 2024, companies already subject to the NFRD have started reporting under the new principles, while from 2025, the same requirements apply to all large EU firms. From 2026, listed small and medium enterprises (SMEs) will also be included, although they may opt for a postponement until 2028. In addition, from 2028, non-EU companies operating in Europe above the applicable thresholds will fall within the scope. Ongoing legislative proposals ("Omnibus package") foresee further simplifications and adjustments to these obligations, together with the introduction of a voluntary standard (VSME) designed to support unlisted SMEs in responding to information requests along the value chain.

The European Sustainability Reporting Standards (ESRS), formally adopted by the European Commission on 31 July 2023, are structured in four blocks (European Commission, 2023):

[16] Information on the roadmap that led to the publication of the CSRD can be found on the EFRAG and European Commission websites, both accessed on 03/09/2025:

https://www.efrag.org/en/projects/sustainability-reporting-standards-roadmap/concluded
https://ec.europa.eu/newsroom/fisma/items/707248/en

- The cross-cutting standards, "ESRS 1—General requirements" and "ESRS 2—General disclosures", containing the general definitions and metrics applying to all sustainability matters.
- The environmental topical standards, "ESRS E1—Climate change", "ESRS E2—Pollution", "ESRS E3—Water and marine resources", "ESRS E4—Biodiversity and ecosystems", and "ESRS E5—Resource use and circular economy".
- The social topical standards, "ESRS S1—Own workforce", "ESRS S2—Workers in the value chain", "ESRS S3—Affected communities", and "ESRS S4—Consumers and end-users".
- The governance topical standard, "ESRS G1—Business conduct", which embraces an extended concept of governance (G+).

Additional standards for listed SMEs and for specific sectors, initially foreseen, have been suspended and are currently under review within the 2025 "Omnibus package".

Complying with ESRS means providing information on all relevant sustainability topics in the areas of governance (i.e., processes, controls, and procedures), strategy (i.e., strategy to interact with impacts, risks, and opportunities), impact, risk and opportunity management (i.e., policies and actions), and metrics and targets (i.e., performance measurement) (European Commission, 2023, p. 5/284). Disclosure requirements are organised in sector-agnostic (i.e., applicable to all undertakings) and entity-specific standards (i.e., additional requirements to cover material information not otherwise disclosed). Sector-specific standards were initially envisaged to apply to undertakings within the same industry; however, their development has been postponed and are currently under review within the 2025 "Omnibus package". This structure ensures that companies can comprehensively address all material aspects, while the balance between sector-agnostic, entity-specific, and potential sector-specific disclosures remains subject to ongoing policy developments.

The information to be disclosed is determined according to the "double materiality" principle, encompassing both impact and financial materiality: The former requires disclosure of the positive or negative impacts that the company may generate on people or the environment, while the latter refers to sustainability matters that could have financial effects on the undertaking (European Commission, 2023, pp. 7–12/284).

Under the former EU NFRD, the comply-or-explain approach—mirroring other voluntary sustainability frameworks such as GRI or Integrated Reporting—allowed companies to omit selected topics based on their own materiality assessment. In practice, many undertakings used this flexibility to disclose selectively, which often reduced transparency and comparability (Pizzi, 2022, p. 77; Venturelli et al., 2020). The CSRD and the ESRS were intended to significantly reduce such discretion, also through the introduction of sector-specific standards designed to address key impacts, risks, and opportunities considered relevant across entire industries (European Commission, 2023, p. 6/284). However, as of 2025, the sector-specific standards are under normative revision within the Omnibus package, and the overall orientation is towards a simplification of disclosure requirements.

In addition to this, the short-, medium-, and long-term impacts should be accounted for not only within the perimeter of business activity but also along the value chain (European Commission, 2023, pp. 13–15/284). The requirements for value chain disclosure may indirectly involve a broader set of enterprises in the reporting process than those formally mandated. SMEs supplying large companies subject to ESRS may be asked to provide information about the impacts of their products or services. Through this mechanism, the EU is encouraging a wider adoption of sustainability principles across the economic system, fostering a shared language similar to the one developed for financial accounting. To prevent SMEs from being disproportionately affected by value chain disclosure requests from larger companies subject to ESRS, the EU has promoted the development of a dedicated voluntary standard. The Voluntary Sustainability Reporting Standard for SMEs (VSME) is designed to support micro-, small-, and medium-sized non-listed enterprises in responding to sustainability-related information demands in a proportionate and accessible way.

In December 2024, EFRAG published the VSME (EFRAG, 2024), which is explicitly recalled by the Commission Recommendation (EU) 2025/1710 of 30 July 2025 (European Commission, 2025). This initiative reflects the EU's intention to encourage a more inclusive and gradual transition, aligned with the broader goals of the CSRD and ESRS framework.

Among the other important aspects of the ESRS regulatory framework, it requires undertakings to disclose information in a standardised XBRL-based digital format, enhancing the comparability, accessibility, and usability of reported information.

In addition to that, the requirements for third-party assurance are becoming more defined, with the European Commission that is expected to adopt a dedicated EU standard by 1 October 2026 to clarify the contents of the required limited assurance for report contents. The previously foreseen transition to reasonable assurance by 2028 is no longer mandatory, following the proposals included in the Omnibus package. These developments reflect the dynamic nature of the regulatory landscape and the EU's commitment to refining the sustainability reporting framework over time.

The regulatory architecture introduced by the CSRD and the ESRS places the EU at the forefront of ESG accounting. However, limiting the analysis to a single continent does not fully capture the global dimension of sustainability challenges. In parallel with the European journey, a global standard-setting process has been launched by the IFRS Foundation—one of the most authoritative institutions in financial reporting—which is discussed in the following section.

1.3.2 The International Standardisation Process: A Focus on the International Sustainability Standards Board (ISSB)

The IFRS Foundation is a not-for-profit organisation founded in 2001 to develop global financial standards to contribute to efficient and resilient capital markets. Since then, the standards proposed by IFRS have achieved global consensus, becoming the global language of financial statements all over the world. As shown in the previous paragraphs, a growing demand for reliable and comparable sustainability information has risen, driven by investors, companies, and financial policymakers. Aware of these needs, in 2021, at COP26 in Glasgow, the IFRS Foundation announced the establishment of the International Sustainability Standards Board (ISSB) to prepare a global baseline of sustainability disclosures focused on the needs of investors and the financial markets.[17]

Several standard and guideline setters, united by the scope of producing investor-oriented sustainability disclosures, joined the IFRS call and are now consolidated into the ISSB. These are the cases of the TCFD, Carbon Disclosure Standards Board, and the VRF, formed by the merger of the SASB and the IIRC. This convergence has resulted in a standardisation de facto, which has significantly enhanced the institutional authority and global relevance of the ISSB from the beginning of its activity, positioning it as a central actor in the development of investor-oriented sustainability reporting frameworks (Pizzi, 2022, pp. 84–88).

The approach proposed by ISSB can be used with any generally accepted accounting principle (GAAP) and is in line with one of the International Accounting Standards Board (IASB), the internal articulation of IFRS that establishes financial standards, ensuring coordination and compatibility between the financial and sustainability areas of reporting. Currently, IFRS sustainability standards are composed of two documents published in June 2023:

- IFRS S1—General requirements for disclosure of sustainability-related financial information, about the core content of disclosure to meet the needs of global capital markets.
- IFRS S2—Climate-related disclosures, containing the guidelines for disclosing on climate-related risks and opportunities.

Complying with IFRS S1 and S2 means disclosing information about an entity's sustainability-related risks and opportunities to support capital providers' investment decisions. In line with TCFD architecture, and similarly to the ESRS, the core contents cover an entity's governance, strategy, risk management, and the adopted

[17] More detailed information about the ISSB, reasons for its establishment, and its scopes are available on the IFRS website: https://www.ifrs.org/groups/international-sustainability-standards-board/ accessed on 03/09/2025.

metrics and targets, but are oriented to determine the effects of sustainability-related information on cash flows, access to finance, or cost of capital over the short-, medium- and long term (IFRS, 2023). This vision constitutes the financial dimension of materiality, for which information is material if its omission could influence investors' decisions.

Building on the SASB metrics and KPIs dataset, IFRS S1 requires industry-specific disclosures and extends the requirements also to value chain evaluations. While only general and climate-related standards have been formally issued, other sustainability topics are addressed through references to existing frameworks such as SASB and Integrated Reporting.

Although IFRS standards facilitate the certification of information, they do not obligate companies to provide assurance for their disclosures. The decision to require assurance is left to individual jurisdictions, which may adopt different approaches based on their regulatory context.

1.3.3 Points of Intersection and Difference among the Standards

The two standardisation processes led by EFRAG and ISSB have produced distinct frameworks that now delineate the boundaries of the sustainability accounting landscape. This subsection aims to highlight the increasing efforts towards interoperability and comparability, while acknowledging the persistent divergences that justify the coexistence of two separate approaches.

Both EFRAG and ISSB explicitly recognise each other's positions and the foundational role played by institutions such as GRI and TCFD in shaping the sustainability reporting discipline. This is reflected in several formal initiatives promoting standard interoperability. For instance, the Memorandum of Understanding between GRI and the IFRS Foundation signals the possibility of a comprehensive global reporting system that integrates financial and impact materiality. Similarly, EFRAG and GRI have maintained a consistent partnership throughout the ESRS development, culminating in a joint statement of interoperability according to which entities reporting under ESRS are considered as reporting with reference to the GRI Standards (EFRAG & GRI, 2023).

A major milestone in the convergence process is the Interoperability Guidance, jointly published by EFRAG and the IFRS Foundation in May 2024 (EFRAG & IFRS, 2024). This document demonstrates a high level of alignment between ESRS and ISSB standards, particularly in climate-related disclosures. It provides practical instructions for preparers aiming to comply with both frameworks and reinforces the shared architectural foundation based on governance, strategy, risk management, metrics, and targets.

Initial divergences between the ESRS and ISSB drafts—such as scope, reporting boundaries, and thematic coverage (Giner & Luque-Vílchez, 2022)—have been partially mitigated (Carungu et al., 2025). The ESRS scope, initially broader and more aligned with the SDGs, was recalibrated following consultations with European stakeholders concerned about administrative burden. Similarly, the value chain reporting requirements under the CSRD, once perceived as more extensive, now show convergence with ISSB's Scope 3 emissions disclosures.

However, two core differences remain: target audience and materiality definition. The ISSB adopts a financial materiality lens, focusing on information relevant to investors and capital markets. In contrast, the ESRS framework embraces double materiality, requiring disclosure of both financial consequences and broader societal and environmental effects, even when the latter do not directly affect enterprise value.

These conceptual differences are closely linked, as the target audience defines the primary users of the reports, and information is considered material if it can influence the decisions that primary users make on the basis of those reports (IFRS, 2023, p. 8). Despite alignment efforts, this divergence remains unresolved: ISSB standards continue to focus on investors and shareholders, while ESRS guidelines extend the primary audience to the plurality of stakeholders. As a result, in the former case, sustainability-related information is material if it affects the enterprise value and modifies investors' assessments (i.e., financial materiality); in the latter, financial materiality is complemented by the need to disclose impacts on society and the environment (i.e., impact materiality), even when financial repercussions are indirect or remote (European Commission, 2023, pp. 7–12).

This second approach is referred to as double materiality and combines both financial and impact dimensions. However, the target audience constitutes only one element of the materiality assessment. Indeed, other components concern the direction of the information flow (i.e., inside-out for impact materiality, outside-in for financial materiality, or both for double materiality) and the time horizon that needs to be considered for assessing the relevance of information (Abhayawansa, 2022; Clark, 2021). For this reason, even though the standardisation process has reduced the elements of divergence to the theme of materiality, this concept remains complex and requires further research for understanding the managerial consequences of adopting alternative approaches.

In summary, the historical and institutional journey traced in this chapter reveals a complex and evolving landscape of sustainability reporting. From early voluntary disclosures to the emergence of global standards, the field has grown in scope, sophistication, and regulatory relevance. However, the coexistence of multiple frameworks and the persistent divergence in their scopes, represented by the different definitions of materiality, leave some questions still unanswered:

What should be disclosed, to whom, and why? How do these choices influence corporate strategy, legitimacy, and performance?

Indeed, the definition and operationalisation of materiality shape not only what is reported, but also how organisations perceive and manage their strategy and sustainability-related issues.

This debate is further explored in Chap. 2, where materiality is examined through internal and managerial lenses, with particular attention to its evolving nature. As materiality is deeply embedded in business, societal, and environmental dimensions, it is inherently subject to change over time and across contexts. For this reason, it must be analysed through a systemic and dynamic lens that captures its complexity and interdependencies.

Chapter 2 will deepen this discussion by exploring the academic debate around the theoretical foundations and practical implications of materiality in sustainability accounting. The goal is to understand how materiality can serve as a strategic tool for performance management, guiding organisations towards more meaningful and transformative sustainability practices, beyond compliance and towards change.

References

Abhayawansa, S. (2022). Swimming against the tide: Back to single materiality for sustainability reporting. *Sustainability Accounting, Management and Policy Journal, 13*(6), 1361–1385. https://doi.org/10.1108/SAMPJ-07-2022-0378

Accountancy Europe. (2019). *Interconnected standard setting for corporate reporting*.

Adamo, S., & Di Cagno, N. (2020). *Il bilancio d'esercizio nelle imprese societarie*. Cacucci Editore.

Adams, C. A. (2015). The international integrated reporting council: A call to action. *Critical Perspectives on Accounting, 27*, 23–28. https://doi.org/10.1016/j.cpa.2014.07.001

Adams, C. A., & Abhayawansa, S. (2022). Connecting the COVID-19 pandemic, environmental, social and governance (ESG) investing and calls for 'harmonisation' of sustainability reporting. *Critical Perspectives on Accounting, 82*. https://doi.org/10.1016/j.cpa.2021.102309

Adams, C. A., Alhamood, A., He, X., Tian, J., Wang, L., & Wang, Y. (2022). The development and implementation of GRI standards: Practice and policy issues. In C. A. Adams (Ed.), *Handbook of accounting and sustainability*. Edward Elgar Publishing Ltd.

Adams, C. A., Druckman, P. B., & Picot, R. C. (2020). *Sustainable development goals disclosure (SDGD) recommendations*. https://www.icas.com/professional-resources/sustainability/sustainable-development-goals/feedback-on-

Adams, C. A., & McNicholas, P. (2007). Making a difference: Sustainability reporting, accountability and organisational change. *Accounting, Auditing and Accountability Journal, 20*(3), 382–402. https://doi.org/10.1108/09513570710748553

Alexander, D., Adamo, S., Di Pietra, R., & Fasiello, R. (2017). *The history and tradition of accounting in Italy* (1st ed.). Routledge.

AlWaer, H., Sibley, M., & Lewis, J. (2008). Different stakeholder perceptions of sustainability assessment. *Architectural Science Review, 51*(1), 48–59. https://doi.org/10.3763/asre.2008.5107

Amaduzzi, A. (1949). *Il sistema produttivo dell'impresa nelle condizioni del suo equilibrio e nel suo andamento*. Signorelli.

Ball, A., & Milne, M. (2005). Sustainability and management control. In A. J. Berry, J. Broadbent, & D. Otley (Eds.), *Management control: Theories, issues, and performance* (pp. 314–337). Palgrave Macmillan.

Bebbington, J. (2021). Foundations of environmental accounting. In J. Bebbington, C. Larrinaga, B. O'Dwyer, & I. Thomson (Eds.), *Routledge handbook of environmental accounting* (pp. 17–28). Routledge.

Biermann, F., Kanie, N., & Kim, R. E. (2017). Global governance by goal-setting: The novel approach of the UN sustainable development goals. *Current Opinion in Environmental Sustainability, 26–27*, 26–31). Elsevier B.V. https://doi.org/10.1016/j.cosust.2017.01.010

Boiral, O., & Henri, J. F. (2017). Is sustainability performance comparable? A study of GRI reports of mining organizations. *Business and Society, 56*(2), 283–317. https://doi.org/10.1177/0007650315576134

Boos, A. (2015). Genuine savings as an indicator for "weak" sustainability: Critical survey and possible ways forward in practical measuring. *Sustainability, 7*(4), 4146–4182. https://doi.org/10.3390/su7044146

Borgert, T., Donovan, J. D., Topple, C., & Masli, E. K. (2018). Initiating sustainability assessments: Insights from practice on a procedural perspective. *Environmental Impact Assessment Review, 72*, 99–107. https://doi.org/10.1016/j.eiar.2018.05.012

Broadstock, D. C., Chan, K., Cheng, L. T. W., & Wang, X. (2021). The role of ESG performance during times of financial crisis: Evidence from COVID-19 in China. *Finance Research Letters, 38*, 101716. https://doi.org/10.1016/j.frl.2020.101716

Bugg-Levine, A., & Emerson, J. (2011). Impact investing: Transforming how we make money while making a difference. *Innovations: Technology, Governance, Globalization, 6*(3), 9–18. https://doi.org/10.1162/INOV_a_00077

Buhmann, K. (2016). Public regulators and CSR: The 'social Licence to operate' in recent United Nations instruments on business and human rights and the Juridification of CSR. *Journal of Business Ethics, 136*(4), 699–714. https://doi.org/10.1007/s10551-015-2869-9

Buhr, N., Gray, R., & Milne, M. J. (2014). Histories, rationales, voluntary standards and future prospects for sustainability reporting: CSR, GRI, IIRC and beyond. In *Sustainability accounting and accountability* (pp. 51–71). Routledge.

Busco, C., Consolandi, C., Eccles, R. G., & Sofra, E. (2020). A preliminary analysis of SASB reporting: Disclosure topics, financial relevance, and the financial intensity of ESG materiality. *Journal of Applied Corporate Finance, 32*(2), 117–125. https://doi.org/10.1111/jacf.12411

Busco, C., Frigo, M., Giovannoni, E., Riccaboni, A., & Scapens, R. W. (2005). Beyond compliance: Why integrated governance matters today. *Strategic Finance, 87*(2), 34–43.

Busco, C., Granà, F., & Achilli, G. (2021). Understanding integrated thinking: Evidence from the field, the development of a framework and avenues for future research. *Meditari Accountancy Research, 29*(4), 673–690. https://doi.org/10.1108/MEDAR-04-2021-1263

Calabrese, A., Costa, R., Levialdi Ghiron, N., Tiburzi, L., & Villazon Montalvan, R. A. (2022). Is the private sector becoming cleaner? Assessing the firm's contribution to the 2030 agenda. *Journal of Cleaner Production, 363*. https://doi.org/10.1016/j.jclepro.2022.132324

Carroll, A. B., Brown, J. A., & Buchholtz, A. K. (2017). *Business & society: Ethics, sustainability, and stakeholder management* (10th ed.). South-Western Pub.

Carungu, J., Di Pietra, R., & Molinari, M. (2020). Mandatory vs voluntary exercise on non-financial reporting: Does a normative/coercive isomorphism facilitate an increase in quality? *Meditari Accountancy Research, 29*(3), 449–476. https://doi.org/10.1108/MEDAR-08-2019-0540

Carungu, J., Dimes, R., & Molinari, M. (2025). EFRAG and ISSB: Tensions and opportunities for convergence in the quest for the standardisation of sustainability reporting standards. *Management Decision*. https://doi.org/10.1108/MD-10-2024-2463

Cassandro, P. E. (1982). *Trattato di ragioneria: l'economia delle aziende e il suo controllo*. Cacucci Editore.

Cassandro, P. E. (1989). Sul cosiddetto bilancio sociale dell'impresa. *Rivista Italiana di Ragioneria e di Economia Aziendale*.

Catturi, G. (2010). Tempi storici e tempi aziendali: aspetti metodologici e conseguenze operative. *Spanish Journal of Accounting History*.

Catturi, G. (2021). *Principi di economia aziendale. L'azienda universale. L'idea forza, la morfologia e la fisiologia* (Settima Edizione). CEDAM.

Ceccherelli, A. (1962). *Istituzioni di Ragioneria*. Le Monnier.

Ceccherelli, A. (1967). *Economia aziendale e Amministrazione delle imprese*. Editrice Universitaria.

Chen, L., Feldmann, A., & Tang, O. (2015). The relationship between disclosures of corporate social performance and financial performance: Evidences from GRI reports in manufacturing industry. *International Journal of Production Economics, 170*, 445–456. https://doi.org/10.1016/j.ijpe.2015.04.004

Cho, C. H., Laine, M., Roberts, R. W., & Rodrigue, M. (2015). Organized hypocrisy, organizational façades, and sustainability reporting. *Accounting, Organizations and Society, 40*, 78–94. https://doi.org/10.1016/j.aos.2014.12.003

Christensen, H. B., Hail, L., & Leuz, C. (2021). Mandatory CSR and sustainability reporting: Economic analysis and literature review. *Review of Accounting Studies, 26*(3), 1176–1248. https://doi.org/10.1007/s11142-021-09609-5

Clark, C. E. (2021). How do standard setters define materiality and why does it matter? *Business Ethics, the Environment & Responsibility, 30*(3), 378–391. https://doi.org/10.1111/beer.12351

Coda, V. (2012). Lectio magistralis: Capire e fare il bene dell'azienda. In *Il bene dell'azienda. Scritti in onore di Vittorio Coda* (pp. 1–42). Giuffrè Francis Lefebvre.

Cordazzo, M., Bini, L., & Marzo, G. (2020). Does the EU directive on non-financial information influence the value relevance of ESG disclosure? Italian evidence. *Business Strategy and the Environment, 29*(8), 3470–3483. https://doi.org/10.1002/bse.2589

Coronella, S., Caputo, F., Leopizzi, R., & Venturelli, A. (2018). Corporate social responsibility in Economia Aziendale scholars' theories: A taxonomic perspective. *Meditari Accountancy Research, 26*(4), 640–656. https://doi.org/10.1108/MEDAR-03-2017-0124

Costanza, R., d'Arge, R., de Groot, R., Farber, S., Grasso, M., Hannon, B., Limburg, K., Naeem, S., O'Neill, R. V., Paruelo, J., Raskin, R. G., Sutton, P., & van den Belt, M. (1997). The value of the world's ecosystem services and natural capital. *Nature, 387*(6630), 253–260. https://doi.org/10.1038/387253a0

Cupertino, S., Vitale, G., & Ruggiero, P. (2022). Performance and (non) mandatory disclosure: The moderating role of the directive 2014/95/EU. *Journal of Applied Accounting Research, 23*(1), 163–183. https://doi.org/10.1108/JAAR-04-2021-0115

Daly, H. E., & Farley, J. (2004). *Ecological economics: Principles and applications*. Island Press.

De Micco, P., Rinaldi, L., Vitale, G., Cupertino, S., & Maraghini, M. P. (2020). The challenges of sustainability reporting and their management: The case of Estra. *Meditari Accountancy Research, 29*(3), 430–448. https://doi.org/10.1108/MEDAR-09-2019-0555

de Villiers, C., La Torre, M., & Molinari, M. (2022). The global reporting initiative's (GRI) past, present and future: Critical reflections and a research agenda on sustainability reporting (standard-setting). *Pacific Accounting Review, 34*(5), 728–747. https://doi.org/10.1108/PAR-02-2022-0034

de Villiers, C., Rinaldi, L., & Unerman, J. (2014). Integrated reporting: Insights, gaps and an agenda for future research. *Accounting, Auditing and Accountability Journal, 27*(7), 1042–1067. https://doi.org/10.1108/AAAJ-06-2014-1736

Deegan, C. (2002). Introduction: The legitimising effect of social and environmental disclosures—A theoretical foundation. *Accounting, Auditing & Accountability Journal, 15*(3), 282–311. https://doi.org/10.1108/09513570210435852

Deloitte. (2022). *Osservatorio Nazionale sulla Rendicontazione Non Finanziaria*.

Dennis, I. (2014). *The nature of accounting regulation* (1st ed.). Routledge.

Di Cagno, N., Adamo, S., & Giaccari, F. (2003). *Lezioni di economia aziendale* (Vol. 1). Cacucci Editore.

Di Pietra, R. (2003). Armonizzazione e standardizzazione contabile tra globalizzazione e localismo. In *Atti del Convegno Nazionale dell'Accademia Italiana di Economia Aziendale* (pp. 1461–1484).

Di Pietra, R. (2015). The nature of accounting regulation. *Accounting in Europe, 12*(1), 127–133. https://doi.org/10.1080/17449480.2015.1054114

Di Vaio, A., Syriopoulos, T., Alvino, F., & Palladino, R. (2020). "Integrated thinking and reporting" towards sustainable business models: A concise bibliometric analysis. *Meditari Accountancy Research, 29*(4), 691–719). Emerald Group Holdings Ltd. https://doi.org/10.1108/MEDAR-12-2019-0641

Dimaggio, P. J., & Powell, W. W. (1983). The iron cage revisited: Institutional isomorphism and collective rationality in organizational fields. *American Sociological Review, 48*(2), 147–160. https://doi.org/10.1515/9780691229270-005

Dimes, R., & de Villiers, C. (2020). How management control systems enable and constrain integrated thinking. *Meditari Accountancy Research, 29*(4), 851–872. https://doi.org/10.1108/MEDAR-05-2020-0880

Donat Castelló, L., Gil-González, D., Alvarez-Dardet Diaz, C., & Hernández-Aguado, I. (2010). The environmental millennium development goal: Progress and barriers to its achievement. *Environmental Science & Policy, 13*(2), 154–163. https://doi.org/10.1016/j.envsci.2009.12.001

Du, S., Bhattacharya, C. B., & Sen, S. (2010). Maximizing business returns to corporate social responsibility (CSR): The role of CSR communication. *International Journal of Management Reviews, 12*(1), 8–19. https://doi.org/10.1111/j.1468-2370.2009.00276.x

Dumay, J., Bernardi, C., Guthrie, J., & La Torre, M. (2017). Barriers to implementing the international integrated reporting framework a contemporary academic perspective. *Meditari Accountancy Research, 25*(4), 461–480. https://doi.org/10.1108/MEDAR-05-2017-0150

EFRAG (2024). *EFRAG voluntary sustainability reporting standard for non-listed SMEs (VSME).* Available at https://www.efrag.org/sites/default/files/sites/webpublishing/SiteAssets/VSME%20Standard.pdf

EFRAG, & GRI. (2023). *EFRAG-GRI Joint Statement of Interoperability.*

EFRAG & IFRS. (2024, May). *ESRS–ISSB Standards.* Interoperability Guidance. Available at https://www.efrag.org/sites/default/files/sites/webpublishing/SiteAssets/ESRS-ISSB%20Standards%20Interoperability%20Guidance.pdf

Elkington, J. (1994). Towards the sustainable corporation: Win-win-win business strategies for sustainable development. *California Management Review, 36*(2), 90–100. https://doi.org/10.2307/41165746

Elkington, J. (2018). *25 Years Ago I Coined the Phrase "Triple Bottom Line." Here's why it's time to rethink it.* Harvard Business Review.

Etzion, D., & Ferraro, F. (2010). The role of analogy in the institutionalization of sustainability reporting. *Organization Science, 21*(5), 1092–1107. https://doi.org/10.1287/orsc.1090.0494

European Commission. (2020). *Financing the green transition: The European green Deal investment plan and just transition mechanism.* EU Press.

European Commission. (2023). *COMMISSION DELEGATED REGULATION (EU) 2023/2772 of 31 July 2023 supplementing directive 2013/34/EU of the European Parliament and of the council as regards sustainability reporting standards.*

European Commission. (2024). *Corporate Sustainability Reporting.* Available at: https://Finance.Ec.Europa.Eu/Capital-Markets-Union-and-Financial-Markets/Company-Reporting-and-Auditing/Company-Reporting/Corporate-Sustainability-Reporting_en

European Commission. (2025). COMMISSION RECOMMENDATION (EU) 2025/1710 of 30 July 2025 on a voluntary sustainability reporting standard for small and medium-sized undertakings.

European Union. (2022). *Corporate Sustainability Reporting Directive (CSRD)–2022/2464.*

Fehling, M., Nelson, B. D., & Venkatapuram, S. (2013). Limitations of the millennium development goals: A literature review. *Global Public Health, 8*(10), 1109–1122. https://doi.org/10.1080/17441692.2013.845676

Feng, T., Cummings, L., & Tweedie, D. (2017). Exploring integrated thinking in integrated reporting—An exploratory study in Australia. *Journal of Intellectual Capital, 18*(2), 330–353. https://doi.org/10.1108/JIC-06-2016-0068

Ferrero, G. (1987). *Impresa e management*. Giuffrè.

Flower, J. (2015). The international integrated reporting council: A story of failure. *Critical Perspectives on Accounting, 27*, 1–17. https://doi.org/10.1016/j.cpa.2014.07.002

Forrester, J. W. (1961). *Industrial dynamics*. Massachusetts Institute of Technology (MIT).

Forrester, J. W. (1971). *World dynamics*. Wright-Allen Press.

Forrester, J. W., Low, G. W., & Mass, N. J. (1974). The debate on "world dynamics": A response to Nordhaus. *Policy Sciences, 5*(2), 169.

Fransen, L., & Conzelmann, T. (2015). Fragmented or cohesive transnational private regulation of sustainability standards? A comparative study. *Regulation & Governance, 9*(3), 259–275. https://doi.org/10.1111/rego.12055

Freeman, R. E. (1984). *Strategic management. A stakeholder approach*. Pitman.

Freeman, R. E. (2008). Ending the so-called "Friedman-Freeman" debate. *Business Ethics Quarterly, 18*(2), 162–166.

Friedman, M. (1962). *Capitalism and freedom*. University of Chicago Books.

García-Meca, E., & Martinez-Ferrero, J. (2021). Is SDG reporting substantial or symbolic? An examination of controversial and environmentally sensitive industries. *Journal of Cleaner Production, 298*. https://doi.org/10.1016/j.jclepro.2021.126781

Gatti, L., Seele, P., & Rademacher, L. (2019). Grey zone in—Greenwash out. A review of greenwashing research and implications for the voluntary-mandatory transition of CSR. *International Journal of Corporate Social Responsibility, 4*(1). https://doi.org/10.1186/s40991-019-0044-9

Gibassier, D., Adams, C., & Jérôme, T. (2019). *Integrated reporting and the capitals' diffusion integrated reporting and the capitals' diffusion overview*.

Gibassier, D., Rodrigue, M., & Arjaliès, D. L. (2018). "Integrated reporting is like god: No one has met him, but everybody talks about him": The power of myths in the adoption of management innovations. *Accounting, Auditing and Accountability Journal, 31*(5), 1349–1380. https://doi.org/10.1108/AAAJ-07-2016-2631

Giner, B., & Luque-Vílchez, M. (2022). A commentary on the "new" institutional actors in sustainability reporting standard-setting: A European perspective. *Sustainability Accounting, Management and Policy Journal, 13*(6), 1284–1309. https://doi.org/10.1108/SAMPJ-06-2021-0222

Goolsby, J. B., Cravens, A. E., & Rozance, M. A. (2023). Becoming an actionable scientist: Challenges, competency, and the development of expertise. *Environmental Management, 72*(6), 1128–1145. https://doi.org/10.1007/s00267-023-01863-4

Gray, R. (2010). Is accounting for sustainability actually accounting for sustainability…and how would we know? An exploration of narratives of organisations and the planet. *Accounting, Organizations and Society, 35*(1), 47–62. https://doi.org/10.1016/j.aos.2009.04.006

Gray, R., Owen, D., & Maunders, K. (1987). *Corporate social reporting: Accounting and accountability*. Prentice-Hall International.

Gray, R., Owen, D., & Maunders, K. (1988). Corporate social reporting: Emerging trends in accountability and the social contract. *Accounting, Auditing & Accountability Journal, 1*(1), 6–20.

GRI. (2002). *G2 Sustainability Reporting Guidelines 2002*.

GRI. (2013). *G4 guidelines*.

Guthrie, J., & Parker, L. D. (1989). Corporate social reporting: A rebuttal of legitimacy theory. *Accounting and Business Research, 19*(76), 343–352. https://doi.org/10.1080/00014788.1989.9728863

Gutterman, A. (2021). Sustainability reporting frameworks. *SSRN Electronic Journal*. https://doi.org/10.2139/ssrn.3809288

Hardi, P., & Zdan, T. (2001). *Assessing sustainable development: Principles in practice* (IISD, Ed.).

Herath, R., Senaratne, S., & Gunarathne, N. (2021). Integrated thinking, orchestration of the six capitals and value creation. *Meditari Accountancy Research, 29*(4), 873–907. https://doi.org/10.1108/MEDAR-01-2020-0676

Hillary, R. (1995). Environmental reporting requirements under the EU: Eco-management and audit scheme (EMAS). *The Environmentalist, 15*(4), 293–299. https://doi.org/10.1007/BF01902251

Hummels, H., & Timmer, D. (2004). Investors in Need of social, ethical, and environmental information. *Journal of Business Ethics, 52*(1), 73–84. https://doi.org/10.1023/B:BUSI.0000033108.20321.f5

IFRS. (2023). *IFRS S1 general requirements for disclosure of sustainability-related financial information*.

IIRC. (2013). *Integrated reporting framework*. www.theiirc.org

IIRC. (2021). *International <IR> framework*. www.integratedreporting.org

IPCC. (2023). *IPCC, 2023: Climate change 2023: Synthesis report* (P. Arias, M. Bustamante, I. Elgizouli, G. Flato, M. Howden, C. Méndez-Vallejo, J. J. Pereira, R. Pichs-Madruga, S. K. Rose, Y. Saheb, R. Sánchez Rodríguez, D. Ürge-Vorsatz, C. Xiao, N. Yassaa, J. Romero, J. Kim, E. F. Haites, Y. Jung, R. Stavins, … C. Péan, Eds.). https://doi.org/10.59327/IPCC/AR6-9789291691647.

Jensen, M. C., & Meckling, W. H. (1976). Theory of the firm: Managerial behavior, agency costs and ownership structure. In *Corporate governance* (pp. 77–132). Gower.

Jollands, S., Akroyd, C., & Sawabe, N. (2018). Management controls and pressure groups: The mediation of overflows. *Accounting, Auditing and Accountability Journal, 31*(6), 1644–1667. https://doi.org/10.1108/AAAJ-10-2016-2747

Kaplan, R. S., & Norton, D. P. (2008). Mastering the management system. *Harvard Business Review, 86*(1).

Klarin, T. (2018). The concept of sustainable development: From its beginning to the contemporary issues. *Zagreb International Review of Economics and Business, 21*(1), 67–94. https://doi.org/10.2478/zireb-2018-0005

Kotler, P., & Lee, N. R. (2004). *Corporate social responsibility: Doing the most good for your company and your cause*. John Wiley & Sons.

KPMG. (2017). *The KPMG survey of corporate responsibility reporting 2017*. www.kpmg.com/crreporting

KPMG. (2020). *The time has come: The KPMG survey of sustainability reporting 2020*.

La Torre, M., Sabelfeld, S., Blomkvist, M., & Dumay, J. (2020). Rebuilding trust: Sustainability and non-financial reporting and the European Union regulation. *Meditari Accountancy Research, 28*(5), 701–725. https://doi.org/10.1108/MEDAR-06-2020-0914

Larrinaga, C., & Bebbington, J. (2021). The pre-history of sustainability reporting: A constructivist reading. *Accounting, Auditing and Accountability Journal, 34*(9), 131–150. https://doi.org/10.1108/AAAJ-03-2017-2872

Levy, D. L., Brown, H., & de Jong, M. (2010). The contested politics of corporate governance: The case of the global reporting initiative. *Business and Society, 49*(1), 88–115. https://doi.org/10.1177/0007650309345420

Lovins, A. B., Lovins, H. L., & Hawken, P. (2005). A road map for natural capitalism. In *Understanding business environments* (pp. 250–263). Routledge.

Macias, H. A., & Farfan-Lievano, A. (2017). Integrated reporting as a strategy for firm growth: Multiple case study in Colombia. *Meditari Accountancy Research, 25*(4), 605–628. https://doi.org/10.1108/MEDAR-11-2016-0099

Malafronte, I., & Pereira, J. (2021). Integrated thinking: Measuring the unobservable. *Meditari Accountancy Research, 29*(4), 805–822. https://doi.org/10.1108/MEDAR-12-2019-0640

Meadows, D. H., Meadows, D. L., Randers, J., & Behrens, W. W., III. (1972). *The limits to growth: A report for the Club of Rome's project on the predicament of mankind*. Universe Pub.

Meadows, D. H., Meadows, D., & Randers, J. (1992). *Beyond the limits*. Chelsea Green Publishing.

Meadows, D. H., Randers, J., & Meadows, D. (2004). *Limits to growth: The 30-year update*. Chelsea Green Publishing.

Melloni, G., Stacchezzini, R., & Lai, A. (2016). The tone of business model disclosure: An impression management analysis of the integrated reports. *Journal of Management and Governance, 20*(2), 295–320. https://doi.org/10.1007/s10997-015-9319-z

Merchant, K. A. (1982). The control function of management. *Sloan Management Review, 24*(4).

Milne, M. J., & Gray, R. (2013). W(h)ither ecology? The triple bottom line, the global reporting initiative, and corporate sustainability reporting. *Journal of Business Ethics, 118*(1), 13–29. https://doi.org/10.1007/s10551-012-1543-8

Murray, A., Skene, K., & Haynes, K. (2017). The circular economy: An interdisciplinary exploration of the concept and application in a global context. *Journal of Business Ethics, 140*(3), 369–380. https://doi.org/10.1007/s10551-015-2693-2

Neumayer, E. (2003). *Weak versus strong sustainability. Exploring the limits of two opposing paradigms* (2nd ed.). Edward Elgar Publishing Limited.

Nilsson, M., Chisholm, E., Griggs, D., Howden-Chapman, P., McCollum, D., Messerli, P., Neumann, B., Stevance, A. S., Visbeck, M., & Stafford-Smith, M. (2018). Mapping interactions between the sustainable development goals: Lessons learned and ways forward. *Sustainability Science, 13*(6), 1489–1503. https://doi.org/10.1007/s11625-018-0604-z

Nordhaus, W. D. (1973). World dynamics: Measurement without data. *Source: The Economic Journal, 83*(332), 1156–1183. https://about.jstor.org/terms

O'Dwyer, B., & Unerman, J. (2020). Shifting the focus of sustainability accounting from impacts to risks and dependencies: Researching the transformative potential of TCFD reporting. *Accounting, Auditing and Accountability Journal, 33*(5), 1113–1141. https://doi.org/10.1108/AAAJ-02-2020-4445

Onida, P. (1953). La dimensione e struttura delle aziende e la produttività. *Produttività, 12*.

Pedercini, M., Arquitt, S., Collste, D., & Herren, H. (2019). Harvesting synergy from sustainable development goal interactions. *Proceedings of the National Academy of Sciences of the United States of America, 116*(46), 23021–23038. https://doi.org/10.1073/pnas.1817276116

Pizzi, S. (2022). *La standardizzazione dell'informativa di sostenibilità*. Cacucci Editore.

Pizzi, S., Caputo, A., Corvino, A., & Venturelli, A. (2020). Management research and the UN sustainable development goals (SDGs): A bibliometric investigation and systematic review. *Journal of Cleaner Production, 276*. https://doi.org/10.1016/j.jclepro.2020.124033

Pizzi, S., Principale, S., & de Nuccio, E. (2023). Material sustainability information and reporting standards. Exploring the differences between GRI and SASB. *Meditari Accountancy Research, 31*(6), 1654–1674. https://doi.org/10.1108/MEDAR-11-2021-1486

Ponzanelli, G. (1977). *Lezioni di Ragioneria generale*. Libreria Ticci.

Rashed, A. H., & Shah, A. (2021). The role of private sector in the implementation of sustainable development goals. *Environment, Development and Sustainability, 23*(3), 2931–2948. https://doi.org/10.1007/s10668-020-00718-w

Reed, G., Hendlin, Y., Desikan, A., MacKinney, T., Berman, E., & Goldman, G. T. (2021). The disinformation playbook: How industry manipulates the science-policy process—And how to restore scientific integrity. *Journal of Public Health Policy, 42*(4), 622–634. https://doi.org/10.1057/s41271-021-00318-6

Rockström, J., Steffen, W., Noone, K., Persson, Å., Chapin, F. S., Lambin, E. F., Lenton, T. M., Scheffer, M., Folke, C., Schellnhuber, H. J., Nykvist, B., de Wit, C. A., Hughes, T., van der Leeuw, S., Rodhe, H., Sörlin, S., Snyder, P. K., Costanza, R., Svedin, U., et al. (2009). A safe operating space for humanity. *Nature, 461*(7263), 472–475. https://doi.org/10.1038/461472a

Sachs, J. (2012). From millennium development goals to sustainable development goals. *Lancet, 379*, 2206–2211.

Salvioni, D. (1992). *Il bilancio di esercizio nella comunicazione integrata d'impresa*. Giappichelli Editore.

Santoni, R. (2023). Integrated Reporting come sistema manageriale per raggiungere obiettivi di sviluppo sostenibile: una verifica empirica. *Management Control, 1*, 43–68. https://doi.org/10.3280/MACO2023-001003

Schaltegger, S., Gibassier, D., & Zvezdov, D. (2013). Is environmental management accounting a discipline? A bibliometric literature review. *Meditari Accountancy Research, 21*(1), 4–31. https://doi.org/10.1108/MEDAR-12-2012-0039

Sitnikov, C., & Bocean, C. (2017). The need for harmonization and certification in social responsibility area. In M. Aluchna & S. Idowu (Eds.), *The dynamics of corporate social responsibility. CSR, sustainability, ethics & governance* (pp. 265–283). Springer. https://doi.org/10.1007/978-3-319-39089-5_14

Sommer, B., & Hain, A. (2017). Europe as a green leader? A brief evaluation of both the European Union's climate and energy policy and common agricultural policy. *Culture, Practice & Europeanization, 2*(2), 33–45.

Stoughton, A. M., & Ludema, J. (2012). The driving forces of sustainability. *Journal of Organizational Change Management, 25*(4), 501–517. https://doi.org/10.1108/09534811211239191

Strada, M. (2023). Gli impatti contabili delle tematiche ESG nei bilanci IFRS. In P. Tettamanzi (Ed.), *Sostenibilità, Impresa e Stakeholder* (pp. 217–239). Giuffrè Francis Lefevre.

Stuart, E., & Woodroffe, J. (2016). Leaving no-one behind: Can the sustainable development goals succeed where the millennium development goals lacked? *Gender and Development, 24*(1), 69–81. https://doi.org/10.1080/13552074.2016.1142206

Sukhari, A., & de Villiers, C. (2019). The influence of integrated reporting on business model and strategy disclosures. *Australian Accounting Review, 29*(4), 708–725. https://doi.org/10.1111/auar.12264

Tay, J. S. W., & Parker, R. H. (1990). Measuring international harmonization and standardization. *Abacus, 26*(1), 71–88. https://doi.org/10.1111/j.1467-6281.1990.tb00233.x

TCFD. (2017). *Recommendations of the task force on climate-related financial disclosures*.

Thomson, I. (2015). 'But does sustainability need capitalism or an integrated report' a commentary on 'the international integrated reporting council: A story of failure' by Flower, J. *Critical Perspectives on Accounting, 27*, 18–22. https://doi.org/10.1016/j.cpa.2014.07.003

Unerman, J., Bebbington, J., & O'Dwyer, B. (2018). Corporate reporting and accounting for externalities. *Accounting and Business Research, 48*(5), 497–522. https://doi.org/10.1080/00014788.2018.1470155

United Nations. (1972). Report of the United Nations conference on the human. *Environment, A/CONF.48/14/Rev*, 1.

United Nations. (1992). *Report of the United Nations Conference on Environment and Development*.

United Nations. (2000). *Millenium Declaration A/RES/55/2*.

United Nations. (2015a). *Paris Agreement*.

United Nations. (2015b). *Transforming Our World: The 2030 agenda for sustainable development A/RES/70/1*.

United Nations. (2023a). *Global indicator framework for the Sustainable Development Goals and targets of the 2030 Agenda for Sustainable Development*.

United Nations. (2023b). *Global sustainable development report. Times of crisis, times of change. Science for accelerating transformations to sustainable development*.

United Nations. (2023c). *Progress towards the sustainable development goals: Towards a rescue plan for people and planet report of the secretary-general (Special Edition)*.

van der Waal, J. W. H., & Thijssens, T. (2020). Corporate involvement in sustainable development goals: Exploring the territory. *Journal of Cleaner Production, 252*. https://doi.org/10.1016/j.jclepro.2019.119625

Vandemoortele, J. (2011). The MDG story: Intention denied. *Development and Change, 42*(1), 1–21. https://doi.org/10.1111/j.1467-7660.2010.01678.x

Venturelli, A., Pizzi, S., Caputo, F., & Principale, S. (2020). The revision of nonfinancial reporting directive: A critical lens on the comparability principle. *Business Strategy and the Environment, 29*(8), 3584–3597. https://doi.org/10.1002/bse.2598

Vermiglio, F. (1984). *Il "bilancio sociale" nel quadro evolutivo del sistema d'impresa*. Grafo.

Victor, P., Hanna, S., & Kubursi, A. (1998). How strong is weak sustainability? In S. Faucheux, M. O'Connor, & J. van der Straaten (Eds.), *Sustainable development: Concepts, rationalities and strategies* (Vol. 13, pp. 195–210). Springer. https://doi.org/10.1007/978-94-017-3188-1_12

Wackernagel, M., & Rees, W. (1996). *Our ecological footprint: Reducing human impact on the earth*. New Society Publishers.

Williams, O. (2018). Restorying the purpose of business: An interpretation of the agenda of the UN global compact. *African Journal of Business Ethics, 12*(2). https://doi.org/10.15249/12-2-195

Wiseman, J. (1982). An evaluation of environmental disclosures made in corporate annual reports. *Accounting, Organizations and Society, 7*(1), 53–63. https://doi.org/10.1016/0361-3682(82)90025-3

World Economic Forum. (2020). *Toward Common Metrics and Consistent Reporting of Sustainable Value Creation*. www.weforum.org

Zappa, G. (1927). *Tendenze nuove negli studi di ragioneria*.

Zappa, G. (1956). *Le produzioni nell'economia delle imprese: Tomo primo*. Giuffrè.

Chapter 2
From Materiality to Performance: The Integration of Sustainability Reporting in Managerial Processes

Contents

2.1 Materiality

The term "materiality" is derived from the adjective "material", meaning important or significant. Materiality is used to describe a qualitative aspect of information in terms of the relevance that it has for a person or in a process. Indeed, the definitions of materiality found in accounting guidelines and general usage tend to converge around a common understanding: Materiality refers to the relevance or significance of information in influencing decision-making. In financial reporting, this typically

R. Santoni, *The Dynamics of Materiality*, System Dynamics for Performance Management & Governance 8, https://doi.org/10.1007/978-3-032-19280-6_2

means that the omission or misrepresentation of certain data could affect the judgments or actions of those relying on the information.[1]

These definitions underline how assessing the materiality of a piece of information is not a peculiar process of sustainability accounting. The first appearances of the notion of materiality can be traced back to UK legal cases in the 1860s and at the beginning of the 1900s in the American accounting texts, disciplines from which sustainability accounting has inherited this term (Clark, 2021; Edgley, 2014; Hicks, 1964).

As described, materiality is a relative concept that refers to the influence that a matter may have on the decision-making process of the information user. Being a relative concept, the importance of identifying the targeted user of such information is crucial. This process is quite intuitive in the financial field, where the investor represents the designated information user, though its definition in the literature may vary between the concepts of informed investor, financial statement user, or, more generally, reasonable person (Edgley, 2014, p. 257).

In financial reporting, the challenge of evaluating materiality has historically emerged more from practical difficulties in its application rather than from the identification of the intended audience. Difficulties include the reliance on subjective judgment, the absence of clear guidelines, and the complexity of measurement (Hicks, 1964, pp. 158–160). One of the most debated issues has been the establishment of thresholds to determine which pieces of information warrant disclosure. In practice, accounting and auditing professionals often rely on pragmatic benchmarks, such as considering information material when its impact exceeds 5% of a firm's net income, supported by literature and professional standards (Edgley, 2014, p. 265). While such thresholds are widely accepted for financial metrics (Eilifsen & Messier, 2015), it is far more difficult to apply similar proxies in the context of sustainability accounting, where the multidimensional nature of environmental and social factors challenges quantification and standardisation.

In the field of sustainability reporting, the concept of materiality has been central since the publication of the first GRI guidelines in 2000, even if the term materiality was not explicitly mentioned. Indeed, GRI guidelines addressed "relevance" as one of the most important reporting principles, according to which firms should identify the threshold at which information becomes significant enough to be reported (GRI, 2002). It was in 2013, with the publication of GRI G4 guidelines, that the principle of materiality determination emerged to reflect the organisation's significant economic, environmental and social impacts that could substantively influence the assessments and decisions of stakeholders (GRI, 2013, p. 11). Since that moment, definitions have spread, with every sustainability reporting framework that has

[1] For example, see the definition of materiality definition by Cambridge Dictionary, available at: https://dictionary.cambridge.org/dictionary/english/materiality. Accessed on 04/09/2025; or by Oxford Dictionary of Accounting, available at: https://www.oxfordreference.com/display/10.1093/oi/authority.20110803100139791#:~:text=The%20extent%20to%20which%20an,decision%20making%20of%20its%20users. Accessed on 04/09/2025.

provided its definition to identify which sustainability issues require disclosure according to its guidelines. Examples include:

- IIRC definition: "An Integrated Report should disclose information about matters that substantively affect the organization's ability to create value over the short, medium, and long term" (IIRC, 2021, p. 29).
- GRI more recent definition of impact materiality: "Topics that represent the organization's most significant impacts on the economy, environment, and people, including impacts on their human rights" (GRI, 2022, p. 26).
- IFRS financial materiality approach: "An entity shall disclose material information about the sustainability-related risks and opportunities that could reasonably be expected to affect the entity's prospects. In the context of sustainability-related financial disclosures, information is material if omitting, misstating or obscuring that information could reasonably be expected to influence decisions that primary users of general-purpose financial reports make on the basis of those reports, which include financial statements and sustainability-related financial disclosures and which provide information about a specific reporting entity" (IFRS, 2023, p. 8).
- ESRS double materiality approach: "Double materiality has two dimensions, namely: impact materiality and financial materiality" (European Commission, 2023, p.9/284). "A sustainability matter is material when it meets the criteria defined for impact materiality [...] or financial materiality [...] or both" (p.8/284), with impact materiality that concerns "the undertaking's material actual or potential, positive or negative impacts on people or the environment over the short-, medium- or long-term" (p.10/284), while "a sustainability matter is material from a financial perspective if it triggers or could reasonably be expected to trigger material financial effects on the undertaking" (p.11/284).

Moving from a conceptual to an analytical perspective, the ESRS guidelines offer a structured approach to materiality assessment by distinguishing between impact and financial dimensions. For impact materiality, actual negative impacts are evaluated based on their severity, which is determined by three factors: scale (how serious the impact is), scope (how widespread it is), and irremediable character (the extent to which the impact can be reversed). For potential negative impacts, likelihood is also considered. Positive impacts follow a similar logic, with materiality based on scale and scope for actual impacts, and on scale, scope, and likelihood for potential ones.

From a financial perspective, a sustainability matter is considered material if it generates, or is reasonably expected to generate, significant financial effects on the organisation. These effects may concern development, financial position, performance, cash flows, access to finance, or cost of capital over the short, medium, or long term. The assessment combines the likelihood of occurrence with the potential magnitude of financial consequences.

The distinction between these two approaches is particularly relevant, as highlighted at the end of Chap. 1. The orientation towards either investors or stakeholders—reflected in the adoption of financial or impact/double materiality—remains the key difference between the outcomes of global sustainability standardisation processes. This choice not only determines what information is disclosed in sustainability reports but also shapes which issues are prioritised and how companies define their strategic responses.

In this context, the Integrated Reporting perspective on materiality—focused on factors that influence an organisation's ability to create value—offers a conceptual bridge between external reporting requirements and internal managerial priorities. It reinforces the idea of materiality as a dynamic and systemic concept, one that evolves in response to changing expectations, risks, and dependencies (IIRC, 2021).

The following Sect. 2.1.1 presents selected academic perspectives that highlight the strengths and weaknesses of each approach, while also identifying areas of convergence and potential integration.

2.1.1 Approaches to Materiality: Strengths, Weaknesses, and Points of Contact

As previously discussed, the subjective nature of materiality assessment remains one of the most debated and controversial aspects of sustainability reporting standardisation. This subjectivity has led to confusion in the application of guidelines and increased the risk of misinterpretation by report users (Abhayawansa, 2022; Jørgensen et al., 2022).

These divergences lie at the heart of an ongoing academic debate, which explores the strengths, limitations, and areas of overlap among the various materiality approaches (Clark, 2021; Cooper & Michelon, 2022). The discussion is often polarised, with scholars and practitioners advocating for the predominance of one approach over others based on its alignment with sustainability goals or its usefulness to report users (Adams & Abhayawansa, 2022).

Advocates of the financial materiality approach emphasise the benefits of comparability and consistency, drawing on the established language and logic of financial reporting. This alignment is seen as enhancing the informativeness of stock prices by embedding financially relevant sustainability information (Schiehll & Kolahgar, 2021). Moreover, firms operating under a market-driven logic tend to favour this orientation (Cerbone & Maroun, 2020).

In contrast, proponents of impact and double materiality consider more important the adoption of a broader perspective of sustainability, capable of accounting for the trade-offs between value for the investors and value for the society, promoting dialogue with stakeholders and stimulating their engagement in business activities to legitimise the firm's role (La Torre et al., 2020). This is because the continuous research of dialogue between the company and the affected communities stimulates a shared vision of what sustainability is, aiming to reach continuous and shared improvement of business behaviour (Adams et al., 2021; Puroila & Mäkelä, 2019). In general, the part of the academic community more focused on the role of the firms in the sustainability transition agrees with the double approach to materiality, as evidenced by the critical comments advanced during the public consultations on the IFRS standards drafts (Adams & Mueller, 2022).

Critics of financial materiality argue that it overlooks environmental and social externalities that may not have immediate financial consequences (Abhayawansa, 2022). Some scholars suggest that the IFRS-led push for standardisation reflects a desire to shift control away from multi-stakeholder processes, rather than a genuine response to investor needs (Adams & Abhayawansa, 2022). Furthermore, the investor community itself is far from homogeneous, and a broader, long-term orientation may ultimately serve their interests more effectively (Christensen et al., 2021; Jørgensen et al., 2022). For these reasons, the financial materiality perspective is increasingly viewed as outdated and insufficient to respond to the growing institutional pressure for a sustainability transition. Its short-term orientation tends to neglect the interests of broader stakeholder groups and fails to account for the complex social and environmental factors that influence organisational activities (Adams et al., 2020; Adams & Mueller, 2022).

On the other hand, despite its aim to connect financial and impact perspectives, the double materiality paradigm has faced criticism. Assessing materiality across both investor and stakeholder dimensions is complex and creates practical challenges. In particular, it is often unclear which issues are relevant to investors and which to other stakeholders (Jørgensen et al., 2022). This ambiguity leads to uncertain reporting boundaries, as the intended audience for each issue is not clearly defined (Abhayawansa, 2022). For this reason, scholars perceive the possibility that double materiality may result in a risk management-oriented approach that privileges the financial orientation of information due to the more circumstantial nature and definiteness of the financial materiality formulation (de Villiers et al., 2022; La Torre et al., 2020).

This debate highlights once again the complex nature of sustainability reporting, characterised by a multiplicity of stakes gravitating around the CSR dimension of the firm, the constant evolution of the normative context, and the company expectations around the adoption of the reporting tool. These aspects require a shift in perspective to reach a unitary vision of the dynamic interplay among the diverse and sometimes conflictual methodologies and objectives within the sustainability accounting domain (Baret & Helfrich, 2019).

Reconciling these divergent approaches requires acknowledging that sustainability reporting serves multiple purposes and audiences. Rather than seeking to establish a single dominant model, it is more productive to recognise the contextual strengths and limitations of each approach.

Chapter 1 already introduced some of the perspectives that view financial and impact materiality as complementary, such as those advanced by the Integrated Reporting framework or the concept of global added value (IIRC, 2021; Catturi, 2021). These complementarities are also reflected in the original accountability purpose of sustainability reporting, which calls for organisations to account for the externalities they generate (Unerman et al., 2018).

In line with this viewpoint, Abhayawansa (2022) identifies the original accountability purpose of sustainability reporting as the key element for bridging the differences among the various materiality approaches developed over time, suggesting a "single materiality" perspective. According to this view, single materiality requires information to be material from the perspective of both investors and stakeholders and not from either or both, allowing the possibility to appreciate the interdependencies between the internal, external, environmental, social, and economic dimensions of sustainability (Abhayawansa, 2022, p. 1376). This perspective aims to make organisations accountable for their actions and promote the use of sustainability reports to foster the green transition rather than just portraying positive business efforts, which is in line with the scope of creating long-term value for the enterprise (Kunc et al., 2021). In this way, the definition of materiality is detached from the decision-usefulness of the information for any stakeholder.

A further integrative concept is that of dynamic materiality, which rejects the static separation between financial and impact dimensions. It recognises that issues considered immaterial today may become financially relevant in the future (World Economic Forum, 2020). This perspective encourages a forward-looking orientation, undocking the materiality assessment from the mere consideration of past data (e.g., what important issues happened the last year, how they have been managed, and consequently decide how to report them) and preparing the business to future issues (Edgley et al., 2015). Detaching materiality from simple reporting purposes (Lai et al., 2017) raises questions about the evolution of a sustainability theme over time and the continuous interrelations between impact and financial repercussions (Kuh et al., 2020).

Nevertheless, even considering dynamicity as a standalone concept is inaccurate, since dynamicity is generated by the relationships among system components (Meadows, 2008). For this reason, performing materiality assessments more often (Garst et al., 2022, pp. 81–82) is not enough to address the complex and socio-political nature of the system (Puroila & Mäkelä, 2019). Indeed, stakeholder engagement procedures may be subject to data collection biases, with stakeholders that may not be completely aware of their priorities (Abhayawansa, 2022, p. 1378) and companies that are often called to address the uncertainty of what matters by "making and giving sense" to materiality (Bolt & Tregidga, 2023).

Differently, considering materiality as involved in a complex stream of information transmission, linking stakeholders and market expectations with business decision-making processes, as well as impact and financial performance measurement (Freiberg et al., 2019), is helpful to understand how the success of a sustainability strategy depends on several interconnected decisional nodes. Thereby, materiality assessment marks the starting point of a cyclical sustainability process, which each year reaches a milestone with the publication of the annual disclosure. This ongoing cycle gradually enhances awareness of the role of accounting in supporting the sustainability transition (Rinaldi et al., 2018).

To understand the repercussions that a financial or a double approach to materiality has on management, it is crucial to retrace the implications of the materiality assessment on the internal sustainability management system. To this end, the remainder of this chapter outlines the information and managerial procedures, from materiality assessment to performance measurement and report publication.

2.1.2 The Materiality Assessment Process

Materiality assessment represents the foundational step in the sustainability reporting process, since it determines the type of information that an organisation should disclose. As previously described, although the concept of materiality—as the relevance of a topic—is intuitively understood, its practical determination is complex and varies depending on the reporting framework adopted and the underlying materiality approach (Clark, 2021).

Differences emerge in stakeholder selection (i.e., who should be considered), engagement techniques (i.e., how stakeholders are involved), and information collection and analysis methods (i.e., how relevance is assessed). These discrepancies hinder the development of a standardised assessment process and obscure how materiality-related insights are used in managerial decision-making (Machado et al., 2021; Zhou, 2011).

This is because some firms still perceive materiality analysis only as a reporting step—devoid of strategic implications—required to assess the extent of measures and narratives to include in the sustainability disclosure (Garst et al., 2022). This approach denotes a compliance orientation of materiality, with firms that often rely on lists of standardised sustainability key issues to conform their report to the expectations of a general audience of non-targeted stakeholders (Borgert et al., 2018).

When the materiality assessment steps are not properly disclosed, the level of transparency of the sustainability report decreases, reducing the comprehensibility of how the report contents have been chosen and their role in corporate strategy (e.g., Machado et al., 2021). The discrepancy among guidelines can be considered one of the possible causes of the generalised opacity on topic prioritisation initiatives. This impacts how the alignment between the firm's strategy and stakeholders' interests is generally assessed, with the logical stream of managerial information—starting with materiality assessment and concluding with performance measurement and strategy adjustment—which is not always visible in business practices (Eccles et al., 2020).

However, the end of Sect. 2.1.1 showed how the return to a single materiality idea (Abhayawansa, 2022) or the introduction of a dynamic conceptualisation of materiality (World Economic Forum, 2020) could bridge the divergent logics by focusing on the existing commonalities. A similar bridging-the-gap idea can be found in the "Applied guidelines on how to identify and monitor the evolution of sustainability-related material issues", published by the "Italian Foundation for Business Reporting—Sustainability, Non-Financial and Integrated Reporting" (OIBR, 2022). These guidelines highlight how firms are supposed to obtain and monitor information derived from several sources and combine it into a unique index stating the materiality level of a certain topic. They offer generalised guidance which encompasses all three perspectives of materiality—that is, financial, impact, double—also highlighting the need for a dynamic approach to materiality which accounts for its changing nature over time (OIBR, 2022).

As shown in Fig. 2.1, financial and impact materiality, the two components of the double materiality approach, do not represent two alternative and contrasting definitions because they share common areas in which a topic can be material both from the outside-in (i.e., the financial and internal repercussions of environmental and social issues coming from the outside) and inside-out (i.e., the impacts that business activities have on the external environment and society) perspectives. Over time, the same topic can move across the boundaries of the two definitions, acquiring or losing financial and impact materiality, according to the dynamic materiality idea. Addressing this movement and change of materiality means considering the time factor of sustainability issues, with fast or slow evolutions of topic materiality often driven by scandals, climate change, geopolitical tensions, law provisions, or consumer awareness that may strongly influence the financial materiality of a topic after a certain period.

Figure 2.1 captures dynamic materiality through the blue arrows and points: Each point, based on its position, indicates a possible materiality state at a given time; the two most external points (on the left and right) represent topics that lie outside the materiality boundaries and are not currently considered material. However, the blue arrows suggest that in future assessments, these topics could become material, entering the realm of financial or impact materiality. Similarly, the three points within the figure represent topics that are currently considered material from different perspectives: The left point reflects financial materiality, the right point pertains to impact materiality, and the centre point holds both financial and impact materiality. The dynamic nature of materiality indicates that these states are not fixed, with the financial and impact significance expected to evolve, especially as many sustainability issues will predictably gain financial relevance.

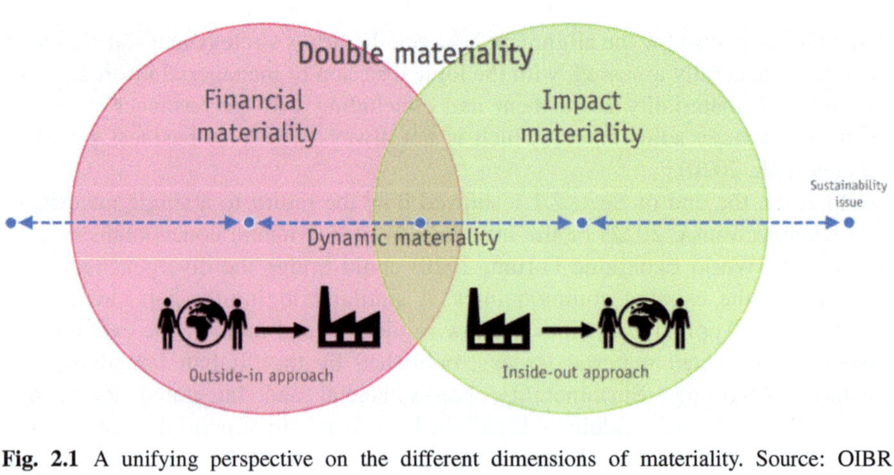

Fig. 2.1 A unifying perspective on the different dimensions of materiality. Source: OIBR (2022, p. 10)

The scope of OIBR materiality guidelines is to fill the operationalisation gap that concerns the materiality assessment (i.e., determinants of materiality) and the related governance processes (i.e., implications of materiality) (OIBR, 2022, p. 4). The creation of adequate governance structures is a prerequisite to integrating materiality judgments into the managerial processes (Eccles & Youmans, 2016). The involvement of top management and interdepartmental governance bodies (OIBR, 2022, p. 17) ensures that materiality assessment is not constrained to a mere compliance procedure, thanks to the diffusion of the awareness of the interdependencies between sustainability issues and business success.

To gain insights on the general importance of a sustainability topic, several sources of evidence can be analysed, such as the disclosures of other companies, existing regulations, sustainability guidelines, media coverage, academic research, and investors' and stakeholders' average interests (OIBR, 2022, p. 25). Though this first step is necessary to understand the general importance of a topic on a global or national scale, it is detached from the business dimension. Therefore, it is inadequate to determine the firm-specific materiality level.

To assign internal significance to the external information, the subsequent steps of the materiality assessment refer to the combination of three aspects: the strategic, impact, and financial relevance of each topic (OIBR, 2022, pp. 26–32). Strategic materiality is obtained by evaluating sustainability-related priorities according to the business perspective, represented by the company's mission, vision, scope, system of values, and managerial beliefs (OIBR, 2022, p. 26). Impact materiality is the result of the analysis of stakeholders'[2] needs and expectations on the business commitment to a particular sustainability topic. It is assessed through surveys, customer service procedures, interviews, workshops, conferences, and direct contact moments organised with customers, employees, institutions, strategic partners, local communities, NGOs, shareholders, and suppliers (OIBR, 2022, p. 29). Financial materiality measures the probability and magnitude of a certain occurrence in terms of financial positive or negative consequences (OIBR, 2022, p. 27).

This process is summarised in Fig. 2.2, where the newly assessed internal materiality influences the strategic dimension of materiality, reorienting business priorities and impacting the company's system of values (Eccles & Youmans, 2016; Lai et al., 2017).

The organisation proceeds to quantify the results of the assessment according to the financial, impact, or double approach adopted. The obtained topic prioritisation is the final output of the materiality assessment process. From the organisations' perspective, this materiality index can be defined as the "internal materiality", that is, the relevance of a sustainability topic for the governance and top management

[2] As shown in Fig. 2.1, the relationship between impact and financial materiality is dynamic and interconnected. Therefore, strictly distinguishing between stakeholders and shareholders may not be necessary. Instead, it is more beneficial to adopt a comprehensive view, where shareholders are considered an integral part of the broader stakeholder group. Even the most impact-focused strategies should take into account the needs of shareholders, as their interests are intertwined with the overall sustainability goals.

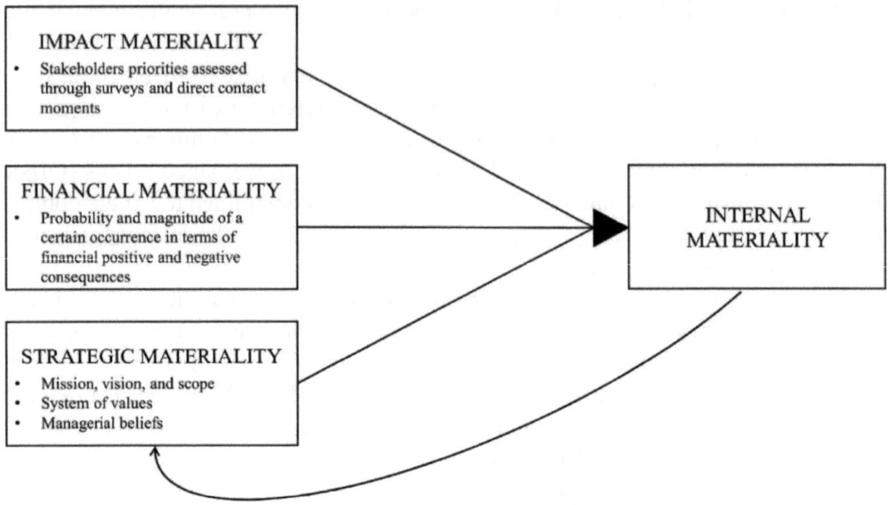

Fig. 2.2 The determinants of internal materiality. Source: Author's elaboration

components. Before each accounting period, the materiality assessment procedure should be performed for all the sustainability issues and disclosed in the organisation's sustainability report. It is possible to locate the results on a scale of measurement (e.g., a matrix) and determine the issues' relative weights in the company's disclosure, decision-making processes, or other business practices (OIBR, 2022, p. 31).

The previously introduced concept of dynamic materiality illustrates how the combination of impact, financial, and strategic dimensions is not static, but subject to continuous change over time.

However, viewing materiality as something that evolves only through occasional, standardised procedures—such as annual assessments and stakeholder engagement—offers a limited perspective. This approach risks reducing dynamic materiality to a discrete-time phenomenon. Moreover, its effectiveness is compromised if companies fail to conduct the assessment annually (as is the case for 27% of the sample in the OIBR study, 2022, p. 33), or if they simply add some minor updates without conducting the entire procedure of stakeholder engagement (Garst et al., 2022, p. 81).

Locating dynamic materiality within a continuous time span of analysis means acknowledging the continuously changing nature of organisations (Tsoukas & Chia, 2002). The level of internal materiality determined by the materiality assessment is promptly challenged by events, with companies dynamically adjusting their behaviour to reach an equilibrium between internal and external pressures and beliefs (Bansal & DesJardine, 2014). Over time, internal materiality becomes embedded in the cultural dimension of the organisation, shaping its system of values. On one hand, this accumulation reflects the company's commitment and predisposition to address sustainability issues (Eccles et al., 2014); on the other, it may lead to path dependency or resistance to change, especially in business-as-usual contexts (Arthur, 1989; Teece et al., 1997).

This feedback process is visually represented by the arrow in Fig. 2.2, which links internal materiality to strategic materiality. The closed loop highlights how materiality assessment functions as a managerial and strategic construct. For example, delays in conducting the assessment translate into delays in strategic adjustments and hinder the transition towards a sustainable business model.

However, adopting a systemic perspective on materiality requires more than acknowledging its influence on internal beliefs, priorities, and orientation. It calls for a deeper investigation into the meanings that organisations attribute to the entire reporting process, positioning materiality as one component of a broader managerial puzzle. To this end, Section 2.2 explores the academic literature on the relationship between sustainability reporting and strategy, with a focus on performance management.

2.2 From Materiality to Performance Management

2.2.1 Materiality, Strategy, and Forward-Looking Orientation

This section explores how materiality assessment informs long-term business strategy and supports a forward-looking orientation in sustainability reporting (Kunc et al., 2021; Tirado-Valencia et al., 2021).

Understanding the dynamic nature of materiality, the constant change and evolution of stakeholders' requirements, perceptions, and desires is critical to understanding the role that an integrated and dynamic approach to materiality can play in the business context (Garst et al., 2022, pp. 81–82). To this point, the idea of developing a managerial topic prioritisation process based on the results of materiality analysis has already been developed and allows firms to set incremental materiality thresholds to substantially align their performance to stakeholders' and shareholders' needs (Calabrese et al., 2016).

Indeed, attributing strategic relevance to the materiality assessment improves the role of sustainability accounting from a passive disclosure to a comprehensive tool to support the value creation process, integrating the sustainability measures into business models, performance and risk management procedures, thereby promoting organisational change (Cerbone & Maroun, 2020; Crovini et al., 2022; Molinari et al., 2021). The benefits of stakeholders' involvement (Torelli et al., 2020a) can be found in the collection of external judgements from people impacted by the firms' activities (Manetti, 2011) that help the business prioritise the sustainability issues (Ferrero-Ferrero et al., 2018) and align the performance measurement procedures (Zarzycka & Krasodomska, 2021).

Also, the normative orientation, expressed by the due diligence approach of the European reporting guidelines, identifies the materiality assessment as a critical step to improve a firm's sustainable behaviour. Sustainability due diligence encompasses the processes through which a firm identifies, prevents, mitigates, and accounts for its actual and potential negative impacts on people and the environment. Such a process shifts the repercussions of the materiality assessment from a backward orientation to a forward-looking one, thanks to the identification of potential sustainability risks and opportunities and the alignment of the governance structure (e.g., informative system, incentive schemes), strategy, and business model (European Commission, 2023).

In line with these approaches, the literature keen on the strategic aspects of business life agrees with the importance of aligning the declared CSR commitment to the business model formulation and risk management conduct (Font et al., 2016). The scope is to overcome the simple legitimation and defensive approach adopted in the majority of sustainability and CSR disclosure to long-term value shared with stakeholders and society in general (Leavy, 2012). This alignment implies embedding the true essence of CSR into corporate strategy, positioning sustainability as a source of long-term competitive advantage (Porter & Kramer, 2006).

The synthesis of all these multi-faceted aspects of CSR alignment with corporate strategy can be found in the unifying concept of integrated thinking[3] which suggests a strong interrelation of the reporting moments, with the materiality assessment used to identify the capitals used or affected during integrated decision-making processes and actions that determine the business behaviour in terms of sustainability performance.

The academic literature agrees on the benefits related to the development of integrated thinking to trigger a virtuous cycle: The publication of sustainability reporting stimulates the creation of integrated thinking (Guthrie et al., 2017), as well as higher levels of integrated thinking facilitate and improve the quality of the entire reporting process (Al-Htaybat & von Alberti-Alhtaybat, 2018). This reinforcing feedback relationship underlines the existence of accumulation processes in which the companies involved in sustainability acquire knowledge and competencies.

Indeed, different stages of reporting maturity have been studied. They depend on the time needed to get engaged with the reporting process, understand its potential and strategic role, attribute decisional weight to the results of materiality analysis, and then use targets and KPIs determined by sustainability challenges to drive decision-making and investment decisions (Farooq & de Villiers, 2019).

These aspects suggest that the adoption of a holistic perspective of business activities (Busco et al., 2021)—which could be articulated in (1) achieving a better understanding of the role of capitals in the creation of long-term value (Herath et al., 2021), (2) integrating environmental and social KPIs in the business model (Busco et al., 2021), and (3) implementing risk management practices at a corporate level (Dimes & de Villiers, 2020; Macias & Farfan-Lievano, 2017)—can stimulate easier

[3] The concept of integrated thinking has already been presented in Sect. 1.2.3 (see IIRC, 2021). To further explore its meaning and managerial implications, the special issue of the journal Meditari Accountancy Research (Volume 29, Issue 4, 2021) provides many points for reflection, including, for example, the articles by Arul et al. (2021); Busco et al. (2024); Kunc et al. (2021); Quarchioni et al. (2021).

processes of change and transition towards sustainable business models (Di Vaio et al., 2020).

To operationalise this holistic perspective, appropriate performance management tools are required; these are introduced in Sect. 2.2.2.

2.2.2 *Performance Management Tools for Sustainability*

Several established frameworks can support the development and implementation of sustainable strategies from a performance management perspective. Among the most important, the ones developed by Kaplan and Norton have a customisable structure that is excellent for understanding the multi-faceted nature of sustainability.

The Balanced Scorecard is introduced as a visual and strategic tool that helps managers translate sustainability goals into measurable objectives across four key dimensions, as depicted in Fig. 2.3. It is a framework that encompasses four key dimensions of business success: learning and growth, internal business process, customer, and financial (Kaplan & Norton, 1992). The model requires managers to identify the critical components on which a company should focus to create value. Within this framework, each of the four perspectives is operationalised through objectives, the first element of each scorecard, which describes the company's strategic goals. As a performance management tool, every objective requires a proper metric to be measured and a target to which to aspire. The actions needed to reach the desired objectives are described under the initiatives column.

While the Balanced Scorecard focuses on measurement, Kaplan and Norton developed a complementary tool, the Strategy Map, to highlight the hierarchical

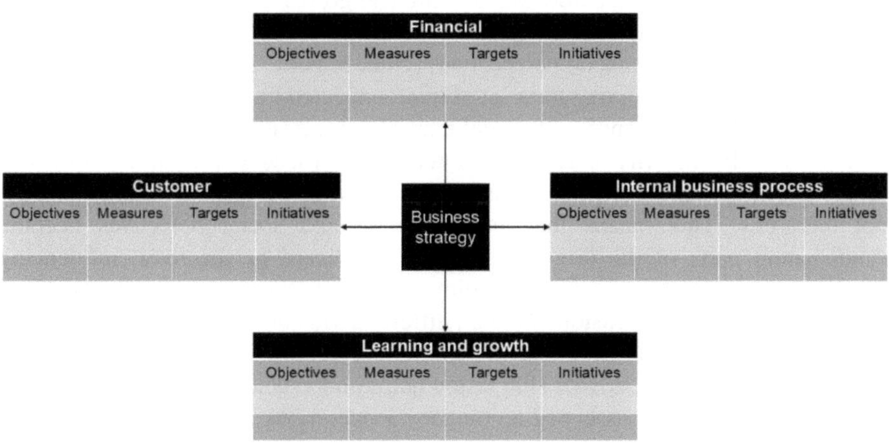

Fig. 2.3 The Balanced Scorecard. Source: Author's elaboration inspired to the Balanced Scorecard by Kaplan and Norton (1996)

relationships among company resources (Kaplan & Norton, 2000). These resources are articulated into the same four perspectives of the Balanced Scorecard and are arranged along an incremental path where knowledge and competencies are considered the basis for business success.

To develop this pyramidal view, the learning and growth dimension, which consists of skills, expertise, culture, and knowledge assets, forms the foundation for all successive steps. The second step is the internal perspective, which focuses on efficiency and value creation. The value a company proposes should meet the expectations of customers, which is the third perspective. Their satisfaction, in turn, generates positive impacts on the financial results of the firm, the fourth and final perspective.

Although the Strategy Map originates from traditional business administration models, its structure is highly customisable and can be adapted to prioritise sustainability goals alongside financial ones (Kaplan & Norton, 2004). When a topic emerges as material through the assessment process, it can be placed at the top of the Strategy Map—whether as a strategic sustainability objective or as a financial target influenced by sustainability concerns.

This positioning allows companies to trace back the determinants of success across the other perspectives of the Balanced Scorecard. By doing so, they can identify the necessary resources and competencies required to achieve the desired outcomes and plan targeted actions accordingly. This logic supports performance improvement on material themes, integrating sustainability into strategic planning and operational execution.

By following this logic, several efforts to adapt the Balanced Scorecard and the Strategy Map to sustainability have been proposed in the academic literature under the label of Sustainability Balanced Scorecard (Hansen & Schaltegger, 2016).

In some cases, environmental and social goals have been added to the original hierarchical structure of the Strategy Map. Sustainability targets are sometimes considered prerequisites (i.e., below the learning and growth perspective) (Figge et al., 2002) and sometimes as ultimate goals at the top of the pyramid (i.e., aligned with or in the place of the financial perspective) (Guix & Font, 2020). Other approaches abandon the hierarchy, opting for a set of interconnected financial and non-financial measures not located along a pyramidal map (Sundin et al., 2010).

However, the majority of academic contributions focus on fully integrating sustainability targets and metrics into the business strategy while maintaining the original hierarchy. This means that environmental and social objectives become fundamental parts of the original Balanced Scorecard perspectives, influencing learning and growth (e.g., green capabilities), internal processes (e.g., sustainability assets and innovations), customers (e.g., differentiation and value recognition), and financial (e.g., eco-efficiency and higher market value of products) perspectives (Hansen & Schaltegger, 2016). This vision helps understand how sustainability elements are fundamental assets of a company's operations and the causal relationships in which they are involved.

Nevertheless, even though the Strategy Map describes the stream of value creation from its foundations to its core KPIs, the causality chain is missing critical elements. More dynamics-oriented literature questions the existence of only linear relationships and the absence of a time dimension (see, e.g., Barnabè, 2008). Some objectives in different Balanced Scorecard perspectives may have mutually reinforcing or conflicting impacts that are not captured in the bottom-up causality stream of the Strategy Map. For example, higher financial performance may generate resources to invest in human resource training (a positive top-down effect), or an increase in the number of customers may drain the capacity to offer post-sale assistance (a trade-off) (Capelo & Dias, 2009).

Regarding the absence of a time dimension, investing in learning and growth is essential for long-term success. However, the impact on customer and financial perspectives may be significantly delayed. This delay may cause managers to lose trust in the Balanced Scorecard's potential, as expected results may not appear in the short term.

To solve these issues, integrating the above-mentioned managerial tools with the dynamicity and the time dimension is crucial for capturing those feedback mechanisms typical of business dynamics. The dynamic performance management (DPM) framework offers a significant contribution in this direction (Bianchi, 2016).

The DPM framework builds on System Dynamics principles and causal modelling to capture feedback loops and time delays in performance outcomes (Forrester, 1961; Morecroft, 2015). It embeds a measurement system to analyse the feedback mechanisms that link the organisational structure to performance outcomes, supporting performance management procedures (Merchant & Van der Stede, 2007). To achieve this, the system's structure—depicted as a network of stocks that are changed by their corresponding flows—affects the system's performance, which is represented by performance outcomes (Sterman, 2000; Vignieri, 2022).

The system's causal structure is framed in three layers: strategic resources, performance drivers, and end-results. Similarly to what occurs in the hierarchical structure of a Strategy Map, the endowment of underlying strategic resources, which could be represented by human competencies or the Integrated Reporting capitals, improves some critical success factors, or performance drivers, which could be exemplified by the time to deliver a product or a service. A positive change in these drivers stimulates a positive impact on outputs (e.g., products delivered) or outcomes (e.g., higher revenues), which are the main objects of managerial performance measurement.

While this simplified causal structure could be seen as a different way of representing and labelling a Strategy Map, the core benefits of DPM models can be identified in the feedback mechanisms they highlight. Every end-result can impact strategic resource endowments, changing the state of the system. For example, material flows of resources change a firm's original asset endowment after every market transaction, while information flows continuously change competencies and skills through knowledge development processes.

These feedback mechanisms add dynamism to the system, and since every stock takes time to accumulate material and immaterial information, they capture the fundamental means-end relationships at the core of managerial decision-making (Merchant, 1982) while overcoming the limitations of Balanced Scorecard-based approaches (Barnabè, 2008). Adopting this perspective is critical when sustainability topics become particularly relevant for a firm (i.e., when the internal materiality level of a topic grows), and managers are asked to develop a strategy to tackle the issue and govern the related performance (Cosma et al., 2022).

To build a DPM system, a company may start by identifying the desired final outcomes—those placed at the top of the Strategy Map—and determining the target KPIs it aims to measure. Then by proceeding backward, the company should identify the key success factors (performance drivers) required to achieve those outcomes, along with the strategic resources needed to influence them. This process enables the organisations to allocate the necessary tangible and intangible assets to effectively address sustainability challenges.

Importantly, the materiality assessment can serve as a starting point for this process. By engaging with stakeholders and analysing their expectations, the company can understand which outcomes are considered most relevant by the intended audience of the sustainability report. These insights allow the organisation to align its entire performance management system with the priorities that emerge from the materiality analysis, ensuring that strategic planning, resource allocation, and KPI selection are all oriented towards the creation of long-term sustainable value.

As an example, imagine that pollution from the production process emerges as a material topic for a company. Managers decide to monitor CO_2 emissions as a KPI for this topic and set a 10% reduction target for the following year. Product carbon intensity—defined as the amount of CO_2 emissions per unit of product manufactured—emerges as a key performance driver. To influence this driver, the company needs to allocate resources towards tangible assets, such as more energy-efficient machinery, and intangible assets, such as the expertise of engineers trained in sustainable production processes. These resources are directly linked to the company's ability to lower its product carbon intensity and, consequently, achieve its overall CO_2 reduction goal.

As previously stated, the peculiarity of the DPM model is that end-results are not necessarily the conclusion of a single path. Indeed, the CO_2 reduction achieved can foster new benefits that create a virtuous cycle. This reduction, for instance, may lead to the creation of a sustainability culture within the company, an immaterial strategic resource that could then foster further CO_2 reductions by leveraging performance paths alternative to industrial product carbon intensity. It can also create new market opportunities by attracting a new segment of customers who value a cleaner product. Furthermore, better ESG performance can improve the terms of bank loans and lead to tax reductions or incentives for greener production. For example, the company may need fewer carbon credits to offset its total CO_2 emissions, making new resources available for alternative investments.

This process exemplifies the benefits of a system-based managerial approach rooted in a materiality assessment. It does not invalidate the benefits of Sustainability Balanced Scorecard–based tools, since the concepts of "perspectives" (e.g., Balanced Scorecard's learning and growth or internal processes), "strategic resources" (the DPM nomenclature), or "capitals" (the Integrated Reporting perspective) substantially refer to the same contents. The key is to project the consequences of the materiality assessment into the future rather than concluding the analysis with the objective of reducing CO_2 emissions. This allows for an analysis of how the outcome changes the company's structure and business model.

In this way, the annual sustainability report should be seen not as a conclusion, but as a starting point for continuous strategic refinement and organisational learning. The company's original stock of resources and capabilities continuously evolves (Frankel, 1962; Yusliza et al., 2020), and the gap between desired and achieved performance becomes the input for renewing the strategy (Merchant, 1982).

By integrating materiality assessment into the broader performance management system, companies can move beyond symbolic compliance and embrace sustainability as a dynamic managerial process. This perspective reframes reporting not as a static exercise but as a feedback-driven mechanism that supports long-term value creation, stakeholder engagement, and strategic adaptability.

Having explored the internal and managerial dimensions of sustainability, the next section turns to the disclosure process itself, examining its theoretical foundations, communicative functions, and external repercussions.

2.3 Theoretical Foundations of Sustainability Reporting and the Role of Materiality

2.3.1 Theories Explaining the Sustainability Reporting Process

The managerial tools discussed in Sect. 2.2 were originally conceived to enhance financial performance, typically through clear and measurable goals. As sustainability has become increasingly central to business practices, these tools have been progressively adapted to incorporate environmental, social, and governance objectives (see Hansen & Schaltegger, 2016). However, unlike financial metrics such as profit, sustainability does not offer a single, universally accepted measure of success. This makes it particularly challenging to define a hierarchy of desired outcomes and to structure a coherent performance management system around them.

Managing sustainability, therefore, goes beyond simply applying existing tools with a "sustainable" label. It requires a deeper analysis of the organisational context and a rethinking of how performance is measured, managed, and communicated. In this perspective, the final value of a single indicator is not sufficient on its own; what truly matters is the role that performance measurement, management, and disclosure processes play within the broader strategic framework.

This idea is also reflected in the function of the sustainability report itself, which primarily serves to summarise the company's diverse efforts and initiatives across ESG dimensions. It communicates performance to stakeholders and shareholders, but often without a clear and shared understanding of how these results are interpreted or used (OIBR, 2022).

To fully comprehend the role of sustainability reporting, it is necessary to explore the underlying motivations that drive companies to disclose non-financial information. Reporting is not a neutral or purely technical activity; it reflects strategic choices, organisational priorities, and the way companies are positioned in their institutional and stakeholder environments.

This section investigates the reasons why companies engage in sustainability reporting, with a particular focus on how the process of selecting and prioritising information, often guided by the materiality assessment, shapes managerial perceptions and disclosure practices.

Over time, various theoretical frameworks have been developed to explain these dynamics (see, among others, Abhayawansa, 2022; De Micco et al., 2020; Hahn & Kühnen, 2013; Larrinaga, 2007; O'Donovan, 2002). Among these, legitimacy, institutional, and stakeholder theories have emerged as particularly influential in understanding the adoption and evolution of sustainability reporting (Fernando & Lawrence, 2014).

Table 2.1 integrates these three theories with additional reading keys, offering a synthesis of the main theoretical contributions. It is important to note that these frameworks are not mutually exclusive. Indeed, they provide complementary lenses through which to interpret the multi-faceted nature of sustainability reporting. Each theory highlights different drivers, mechanisms, and consequences of disclosure practices, and their combined use can enrich both academic analysis and practical understanding. Recognising this complementarity helps mitigate the risk of theoretical fragmentation and supports a more holistic view of sustainability accounting.

Building on these theoretical foundations, Section 2.3.2 explores how the concept of materiality operates within different reporting orientations. In particular, it investigates whether materiality serves primarily to internalise sustainability information for strategic decision-making, or to externalise communication and shape stakeholder perceptions.

2.3.2 The Role of Materiality: Internalising Information or Externalising Communication?

Within this complex and multi-faceted context, can the debate around materiality be reduced to its financial and impact dimensions alone? Or should the topic prioritisation process account for the broader meaning that a company attributes to sustainability (Remmer & Gilbert, 2019)?

In the traditional view, financial materiality is framed as an outside-in approach, focusing on how sustainability issues affect the company. In contrast, impact materiality adopts an inside-out perspective, emphasising the effects of corporate activities on society and the environment (EFRAG, 2021, p. 34; OIBR, 2022, p. 10).

However, the internal or external orientation of the informative flow does not always align precisely with these definitions. It often depends on how companies interpret and use the materiality assessment, as highlighted by the theoretical frameworks that explain corporate behaviour. For instance, Beske et al. (2020) compare

Table 2.1 Key theoretical frameworks in sustainability reporting literature

Theory's name	Theoretical insights and key scientific references
Institutional theory	Institutional theory suggests that sustainability reporting is driven by external pressures from regulatory bodies, industry norms, or societal expectations (Larrinaga, 2007). Over time, sustainability reporting becomes institutionalised as a standard practice that companies are expected to follow (Meyer & Rowan, 1977), often leading to common reporting structures and dominant guidelines (de Villiers & Alexander, 2014).
Stakeholder theory	Stakeholder theory is a specification of the broader agency theory where company managers (i.e., agents) respond to the accountability needs of a plurality of stakeholders (i.e., principals) and not just shareholders (Freeman, 2008). Companies disclose information to address stakeholder pressures (Fortanier et al., 2011), foster engagement, and align their activities with issues considered most material (Fasan & Mio, 2017).
Legitimacy theory	Closely related to stakeholder theory (De Micco et al., 2020), legitimacy theory argues that companies report on their sustainability efforts to gain legitimacy in the eyes of society and obtain a social "license to operate" (Deegan, 2002). In this view, sustainability reporting becomes crucial when societal perceptions of a company are negative (Hahn & Kühnen, 2013).
Signalling theory	Signalling theory addresses situations of information asymmetry, where companies use sustainability reporting as a signal to convey their commitment to responsible practices and to reduce uncertainty between companies and their stakeholders (Connelly et al., 2011; Hahn & Kühnen, 2013). However, signalling theory is often used in combination with greenwashing or impression management theories (Uyar et al., 2020), since reporting can be used opportunistically to enhance reputation (Diouf & Boiral, 2017).
Actor-network theory	Actor-network theory highlights the relational nature of sustainability, where human and non-human actors are involved in continuous interactions, shaping sustainability practices (Callon, 1984). It could be used to frame stakeholders as representatives of nature, giving voice to non-human actors (Callon, 1984; Quattrone, 2022).
Stewardship theory	Stewardship theory (Davis et al., 1997) emphasises that managers act as stewards of both the company and society's interests and are responsible for the effective and ethical use of resources. Core values such as transparency, integrity, trust, and collaboration are considered essential in guiding organisational behaviour and reporting practices (Contrafatto, 2014).
Learning-based theory	Learning-based theory views sustainability reporting as a tool for organisational learning, connecting it to cognitive processes that can drive organisational change (Albrecht et al., 2007). The process of collecting and disclosing sustainability data helps companies reflect on their practices, set new goals, and gradually integrate sustainability into their business strategies, promoting continuous improvement (Farooq & de Villiers, 2019), in line with the learning-based approach to research (Al-Htaybat & von Alberti-Alhtaybat, 2018; Busco et al., 2021).
Ecological modernisation theory	Ecological modernisation theory suggests that companies view sustainability as compatible with economic growth and the existing economic and political systems (Spaargaren & Mol, 1992). Innovation and technological advancements are crucial to fostering efficiency and promoting economic and environmental progress (Weber & Weber, 2020), with sustainability reporting that should explain the company's strategy and investments.

Source: Author's elaboration inspired by "Table 2.1—Literature streams on sustainability reporting" (De Micco et al., 2020, pp. 10–11)

companies oriented towards stakeholders with those whose reports could be categorised as legitimacy-driven. In the first case—although the stakeholder orientation suggests an inside-out impact approach—companies actively manage competing stakeholder interests, developing detailed procedures to identify, prioritise, and internalise sustainability issues, resulting in an outside-in orientated approach guided by strategic integration. On the contrary, companies looking for legitimisation, intuitively more oriented to an outside-in financial approach, may be driven by the objective to comply with social norms and expectations, resulting in a materiality assessment that is often superficial and externally oriented.

Other studies adopting institutional theory describe isomorphism as a tactic used by companies to ensure survival and stability. This often leads to disclosure practices that conform to external expectations without necessarily advancing internal sustainability management (Carungu et al., 2020; Fernando & Lawrence, 2014).

Therefore, it is not the formal classification of materiality that determines the orientation of the informative flow. What truly matters is how the information is used within the organisation and the underlying purpose of the reporting process. Whether aimed at strategic integration, stakeholder engagement, or reputational management, these intentions shape the way materiality is operationalised, regardless of the framework adopted (Tregidga et al., 2018).

For this reason, it is useful to classify the eight theories presented in Table 2.1 based on their tendency to either stimulate internal strategic change (outside-in) or enhance the external perception of corporate behaviour (inside-out), recognising that these orientations may not always correspond directly to the formal definitions of financial and impact materiality.

Theories that predominantly drive internal and strategic change include:

- *Stakeholder* approaches: Stakeholders exert pressure and can shape company behaviour; materiality is useful to align companies' actions with stakeholders' priorities, promoting accountability and co-defining strategy.
- *Actor-network theory* approaches: The interconnectedness of human and non-human actors within the company's network encourages companies to consider their active role in the sustainability definition; the most relevant topics become part of internal processes.
- *Stewardship* approaches: Managers are framed as stewards who are responsible for the long-term health of the company, its stakeholders, and society. This view promotes internal ethical responsibility, with materiality assessment that should foster transparency, trust, and integrity while guiding decision-making and sustainability practices.
- *Learning* approaches: Sustainability reporting is a tool for organisational learning and transformation, with companies that gain insights, adapt, and make internal improvements to embed most material sustainability topics into business strategy over time.
- *Ecological* modernisation approaches: Companies adopting this approach focus on internal processes to improve efficiency and integrate environmental concerns into their operations. Due to the importance of investments and technological

innovations, materiality assessment is crucial to efficiently allocate available resources.

In these theories, sustainability is regarded as an integral part of the company's internal culture, decision-making processes, and operational framework, promoting long-term strategic alignment, internal development, and expected improvement in financial (Khan et al., 2016) and ESG performance (Gerwanski et al., 2019). Materiality assessment is closely linked to strategy definition and revision (Lai et al., 2017), generating a successful re-orientation of the business strategy (Cosma et al., 2022).

Differently, theories with a stronger external orientation include:

- *Institutional* approaches: Firms are driven by external pressures from regulations, norms, and industry standards. Their sustainability practices are often aligned with what is considered acceptable by institutions, regulators, competitors, and the public, ensuring market compliance.
- *Legitimacy* approaches: These companies report on sustainability to gain or maintain a social license to operate, to show society that their practices are acceptable, and to improve their public image.
- *Signalling* approaches: Companies engage in sustainability reporting to signal their commitment to ethical and responsible behaviour, especially in situations where there is information asymmetry between them and the public. By highlighting positive actions, they aim to enhance their reputation and build trust with stakeholders.

Here, materiality is primarily used to determine which topics should receive greater disclosure emphasis, based on their perceived relevance (Calabrese et al., 2016), rather than to support strategic decision-making. This reflects a perspective closely aligned with the traditional concept of materiality in financial accounting, where the focus lies on the usefulness of information for external users rather than its managerial implications.

While this classification of theoretical orientations helps clarify the underlying logic of different reporting approaches, it is important to recognise that, in practice, companies often operate along a continuum between internal strategic integration and external perception management. Materiality assessments may serve both purposes simultaneously, depending on organisational priorities, stakeholder expectations, and contextual pressures.

However, this duality can give rise to tensions between strategic intent and communicative practice. Companies may genuinely aim to integrate sustainability into their core strategy, but at the same time feel obliged to meet external expectations through persuasive or symbolic communication. This becomes particularly problematic when sustainability goals are multiple, diverse, and sometimes conflicting, making it difficult to maintain coherence between what is declared and what is actually pursued.

What happens, for instance, when a topic emerges as material in an area where the company has neither invested nor developed measurement capabilities?

In such cases, the pressure to disclose may lead to superficial or selective reporting, where the narrative does not reflect the underlying organisational reality.

This question points to the risk of misalignment between declared priorities and actual performance and calls for a deeper exploration of the potential space for misleading disclosures and greenwashing, issues that are addressed in Sect. 2.3.3.

2.3.3 The Dark (or Green?) Side of Reporting

The tension between strategic intent and communicative practice, highlighted at the end of Sect. 2.3.2, raises important questions about how companies use sustainability reporting to shape external perceptions. In this context, signalling theory (Connelly et al., 2011) offers a useful lens to interpret corporate disclosure behaviour.

During the reporting process, companies decide which information to communicate, or signal, to external stakeholders, including investors, regulators, and civil society (Clarkson et al., 2008). These stakeholders then interpret the signals and assess their credibility and reliability, forming judgments about the company's sustainability commitment.

Signalling theory has been used to support two separate perspectives on the role of sustainability reporting (De Micco et al., 2020). On one hand, it can be seen as a tool to reduce information asymmetry and promote transparency, helping report users better understand corporate sustainability efforts (Uyar et al., 2020). On the other hand, it can also enable opportunistic behaviour, where companies exploit these asymmetries to enhance their reputation without corresponding improvements in actual performance (de Freitas Netto et al., 2020).

Such opportunism may also emerge in response to conflicting institutional pressures (see Cerbone & Maroun, 2020) or through legitimacy strategies in which the materiality assessment process is not disclosed and remains unverifiable from the outside (Beske et al., 2020). More critically, Cho et al. (2015) argue that neither signalling theory nor impression management theory (often considered extensions of legitimacy theory) fully capture the complexity of corporate disclosure behaviour. Instead, they suggest that organisations may construct inconsistent narratives tailored to different audiences, resulting in a phenomenon defined as "organised hypocrisy".

This concept reflects the gap between declared sustainability commitments and actual practices and highlights the need to investigate how companies use materiality not only to guide strategy but also to manage their sustainability image.

In this context, it is crucial to recognise that materiality assessment, when misused, can become a key driver of distortion in sustainability reporting (Zharfpeykan, 2021). This misuse can take different forms. In some cases, companies may disregard the results of the materiality analysis, choosing instead to highlight areas where performance is stronger (even if those issues are less relevant), thus producing reports misaligned with strategic priorities (Maniora, 2018). In other cases, the materiality assessment itself may be used to pilot the disclosure process, exaggerating achievements on selected topics to gain additional legitimacy from stakeholders (Ferrero-Ferrero et al., 2021).

These behaviours suggest that materiality is not always employed as a tool for strategic alignment, but can be instrumentalised to support reputational goals, especially when the reporting process lacks transparency or when stakeholder engagement is superficial.

The literature offers several terms to describe such tactics, including the previously mentioned impression management (Diouf & Boiral, 2017; Melloni et al., 2016) and organised hypocrisy (Cho et al., 2015). However, they all point to a common issue: the misalignment between actual sustainability performance and the way it is communicated. This phenomenon is widely referred to as greenwashing, which describes the coexistence of poor or limited sustainability performance with overly positive or misleading disclosure practices (Delmas & Burbano, 2011).

Greenwashing can manifest through both explicit behaviours and subtle narrative strategies, often tailored to the company's cultural context (Jones, 2019). It may occur at the corporate or product levels, through executional practices (such as the use of sustainability-evoking elements) or through direct claims that overstate environmental or social achievements (de Freitas Netto et al., 2020).

Although greenwashing is difficult to detect from outside the organisation, its perception by stakeholders can seriously undermine the credibility of the entire sustainability reporting system (La Torre et al., 2020). Misleading communication may initially enhance the company's image, which is often the intended outcome of such practices. However, this short-term reputational gain comes with significant risks. Once greenwashing is perceived, it tends to generate long-term scepticism, particularly when stakeholders feel that the discrepancy between actual performance and communicated attributes is substantial (Torelli et al., 2020b).

This rise in scepticism erodes trust among consumers and stakeholders, ultimately harming the company's reputation and financial performance (Aji & Sutikno, 2015).

The model developed by Calabrese et al. (2015) analyses the relationship between companies and stakeholder trust and satisfaction, identifying three key components that shape stakeholder feedback on corporate social responsibility commitment.

The first component is the performance disclosed by the company through its sustainability report or other communication channels. The second is the perceived commitment, which reflects how stakeholders interpret the disclosed performance. This perception depends on the credibility they attribute to the company: When greenwashing is suspected, perceived commitment declines, undermining both trust and the reliability of the reported information. The third component is the required commitment, which captures stakeholders' expectations and normative demands regarding the company's approach to sustainability.

According to this framework, the adequacy of a sustainability report is not determined solely by the quantity or quality of disclosed performance, but also by how credible and aligned that information appears to stakeholders (Calabrese et al., 2019). The concept of stakeholders' satisfaction, defined as the degree to which actual performance matches prior expectations (Oliver, 1980), can be used to summarise the interaction between required, disclosed, and perceived commitment (Calabrese et al., 2015).

Greenwashing can also be closely linked to how companies approach materiality assessment. In particular, materiality can be used not only to highlight favourable social and environmental impacts, but also to obscure unfavourable ones by omitting negative externalities (Zharfpeykan, 2021). In the reporting context, the ambiguity surrounding materiality definitions, also represented by conflicting guidelines, creates space for companies to manipulate the assessment process to serve reputational interests, tailoring reports to emphasise selected topics (Puroila & Mäkelä, 2019).

As a consequence of these activities, it is essential to distinguish between the internal and external dimensions of materiality. Internal materiality refers to the strategic relevance of a topic for the company, influencing planning and resource allocation. External materiality, on the other hand, concerns the prioritisation of topics for disclosure, shaping the narrative presented to stakeholders.

An example can clarify the distinction between internal and external materiality. Suppose a manufacturer of disposable, non-recyclable products in a country with no penalties disincentivising these products. While stakeholders may express concern about pollution, the absence of regulatory or market consequences means the issue lacks financial materiality. The matrix depicted in Table 2.2 analyses this scenario, illustrating how different combinations of reporting purpose (internal vs. external) and materiality approach (financial vs impact) influence disclosure decisions.

Table 2.2 Example matrix: Possible interactions between reporting purpose and materiality approach

Main reasons for reporting and assessing materiality	Internal purpose	External purpose
Financial materiality	A: The issue is not considered material. B: The issue is not reported.	A: The issue is not considered material. B: Possible scenarios: – If external pressure exists, financial materiality may be used to justify omission (a form of greenwashing). – If no pressure exists, the issue is not reported.
Impact/double materiality	A: The issue becomes strategically relevant, prompting investment. B: The company discloses poor performance and outlines improvement actions.	A: The issue is externally material but not internally prioritised. B: The company may: – Disclose poor performance. – Engage in greenwashing by omitting or misrepresenting the issue (e.g., through a biased materiality matrix).

Results are expressed in terms of materiality of the issue (A) and possible reporting scenarios (B).[4]

The matrix highlights three key aspects that shape a company's approach to sustainability based on its materiality assessment. The first concerns how materiality is assessed, that is, through financial analysis or stakeholder engagement techniques. The second relates to the interpretation of the materiality index, specifically whether it is used for internal strategic purposes or external communication. The third involves the response to external pressures: even companies that adopt a purely financial materiality approach must address stakeholder demands for information. In such cases, tactics such as omitting negative externalities, offering narrative justifications, or diverting attention to less critical issues may emerge, practices commonly associated with impression management (see Melloni et al., 2016) and linked to the broader concept of greenwashing.

An important aspect emerges from this reasoning. Internal materiality, while intended to guide strategic decisions and resource allocation, can also influence greenwashing behaviours. This happens when a topic is not prioritised internally (i.e., low internal materiality) but is subject to strong external scrutiny (i.e., high external materiality). In these situations, companies may be tempted to manipulate their reporting, emphasising favourable narratives to meet stakeholder expectations, without addressing the actual impact of their operations.

In this way, it becomes clear that the external perception of a company's sustainability performance is shaped by multiple factors, many of which go beyond the actual contents of corporate disclosures and enter the realm of stakeholder interpretation and judgement (Calabrese et al., 2015).

Therefore, Section 2.4 shifts the focus to stakeholders, examining how their expectations are linked to their perception of materiality and how these expectations can be interpreted as a dynamic benchmark for assessing sustainability performance. For companies, obtaining a licence to operate increasingly depends on their ability to address stakeholder priorities, an idea central to the impact dimension of sustainability (Busco et al., 2024).

2.4 The Impact Dimension of Sustainability

The previous subsection highlighted how stakeholders' levels of satisfaction depend on the comparison of their expectations with a company's performance. This section continues the exploration of the theme by focusing on the determinants of stakeholders' expectations, which shape the relationship between the stakeholders and the company under investigation.

---

[4] A practical example of the bottom-right quadrant is provided by Maniora (2018), who describes a company that refused to publish a negative report. Instead, it adapted the materiality assessment to divert stakeholder attention toward less relevant topics where performance was stronger, thereby obscuring underperformance on more critical issues.

This exploration needs to be rooted in the reconstruction of the overall meaning of the concept of sustainability, described in the historical reconstruction made in the first chapter of this book. There, the discussion pointed out how the concept of sustainability has been shaped over time, from the first alarms on the pollution generated by economic activities to the definitions of sustainability as something that contemporarily meets the needs of environmental, social, and economic dimensions, or those of the present and future generations (Elkington, 1994; WCED, 1987). In addition to this, more measurement-oriented frameworks have identified some planetary boundaries over which the capacity of the environment to restore itself would be permanently damaged (e.g., Rockström et al., 2009). Then, the efforts made by States and supranational entities to establish some common rules and targets to reduce inequalities and keep climate change within a still manageable dimension. This could be the example of the 17 SDGs or of the Paris Agreement, made to keep global warmth below the +1.5 °C threshold. However, whether they are the result of scientific research or political agreement, these objectives may help define sustainability from a global perspective, but they are extremely difficult to translate into objectives at the business level.

Indeed, the concept of sustainability for an entity lacks absolute thresholds or defined targets, since it can significantly vary according to different criteria and perspectives dependent on context-specific aspects (Hoffman & Woody, 2008). A solution to identify levels of performance that can be considered appropriate to be sustainable has not yet emerged, neither with the greater orientation to measurement introduced by the SDGs. This left unsolved the problem of the lack of a sustainable benchmark to which companies could aspire, with the results of existing mathematical techniques considered sometimes narrow and non-comparable (Pazienza et al., 2023), in other cases non-reliable due to divergencies in results (Berg et al., 2022).

In this situation, the impossibility of setting objective and scientifically rooted performance targets to be sustainable makes the context in which companies operate increasingly relevant. Indeed, the external pressure exerted by stakeholders and the society in general becomes a proxy to determine which level of performance can be considered sustainable. Therefore, understanding the determinants of stakeholders' expectations is a critical starting point to set a dynamic sustainability performance benchmark, reflecting context and time peculiarities and allowing the opportunity to expand the time span of the research on materiality implications.

To this point, Table 2.1 outlined several theories explaining how external expectations and pressures impact organisations' activities. Examples could be the accountability demands based on agency relationships (i.e., stakeholder theory, Freeman, 1984) or companies' desire to acquire social licence to operate (i.e., legitimacy theory, Guthrie & Parker, 1989). Supporting such perspectives, extensive literature has dealt with the increasing interests of communities and institutions in business environmental and social performance (Carroll, 1991), both in terms of exerted pressure (Millar & Searcy, 2020) and interest in the company's disclosure and reporting processes (Fernandez-Feijoo et al., 2014; Prado-Lorenzo et al., 2009).

In any case, keeping the viewpoint of a single organisation, the collective influence of external pressures can be consolidated into a single concept of "stakeholders' expectations", which could be considered a benchmark against which performance should be aligned (Freiberg et al., 2019).

Continuing with this reasoning, companies need to achieve and communicate levels of performance adequate to satisfy stakeholders' requirements. In this way, stakeholders' satisfaction, measuring the adequacy of business performance with stakeholders' requirements, can be effectively considered a proxy for determining the success of the company's sustainable strategy. This does not mean that the company is sustainable in absolute terms, since the discussion showed how this cannot be stated objectively. Instead, it set a dynamic and context-specific benchmark, necessary to adequately build performance management systems and integrate sustainability into the company's strategy.

Following this logic, how can this dynamic benchmark be effectively determined and managed?

In the continuously changing environment of sustainable development, the open and dynamic approach of the goal-setting theory helps define the mechanism that shapes the relationship between the importance of an issue and the related expectations (Locke & Latham, 1990). From the business perspective, organisations utilise specific goals to direct their activities towards desired and measurable outcomes. The ambitiousness of these targets is based on the importance of a topic according to the organisation's priorities, and it determines the level of effort that a company should apply. Once defined, goals do not remain static but are embedded in a dynamic performance management system (Moynihan, 2008), which provides a feedback mechanism that ensures the goal adjustment over time to maintain high levels of motivation and calibrate the required efforts.

Determining a goal corresponding to stakeholders' expectations means mediating among different stakeholders' priorities (AlWaer et al., 2008): For example, dealing with institutions, banks, or NGOs' expectations may suggest the use of targets such as respecting legislative measures to contain pollution (e.g., limits to the release of wastewater), improving specific sustainability metrics taken into account by banks (e.g., the one impacting ESG scores and rankings), or avoiding crossing red lines that can cause heavy protests or boycott actions (e.g., respect for animals and biodiversity). Moreover, stakeholders' priorities evolve rapidly, reflecting the social and environmental changing circumstances (Abhayawansa, 2022). In this context, companies should dynamically identify and adapt desirable goals to guide their activities and satisfy stakeholders.

As can be easily understood, the process of investigating stakeholder priorities is the materiality assessment. This allows companies to internalise external expectations and translate them into actionable performance targets.

However, this process is not without significant challenges. Operational hurdles—such as data collection biases, divergencies across the impact, financial, and strategic components of materiality (as detailed in Fig. 2.2, Section 2.1.2), or other execution difficulties—can hinder the effective implementation of the resulting performance management system.

Furthermore, even with a flawless collection of stakeholder priorities, internal misalignment is possible. Decision-makers may prioritise other sustainability elements, often driven by a financial approach or reliance on pre-established strategic goals. Consequently, external stakeholder priorities may diverge from internal targets, even when materiality assessments are performed and integrated into strategy.

Nevertheless, despite these inherent limitations and difficulties, establishing a dynamic performance benchmark that is integrated into core decisional processes represents a crucial advancement for the sustainability accounting discipline. Therefore, while further research is required to resolve the persisting operational bottlenecks, the focus must now shift to the determinants of stakeholder priorities and how these evolve over time.

Similarly to what happens in a performance management system, a control mechanism exists also in the case of stakeholders' expectations. It is related to the psychological relationship that links the sense of satisfaction—derived from the results achieved—with the worry or concern about the original issue. According to stress and coping theory (Lazarus & Folkman, 1984), people continuously look for solutions to events or issues that are perceived as threatening and stressful. In the previous examples on institutions, banks, or NGOs, each stakeholder category made use of different coping efforts to manage and improve stress situations (e.g., law provisions, rankings, protests). The effectiveness of the coping strategy determines the resolution of the issue or the continuation of the stress: A company's adequate answers to stakeholders' pressure reduce stakeholders' sense of stress and anxiety, thus decreasing the cause of pressure. This cause of pressure can be conceptually defined as stakeholders' materiality, thus establishing a coherent terminology for both the internal (to the company) and external dimensions of sustainability.

Stakeholders' materiality represents the significance of an issue to a group of stakeholders: The more material an issue is, the more it is likely to impact stakeholders' decisions, actions, and consequently, expectations. In this way, it constitutes the source of stakeholders' expectations, and, according to stress and coping theory, it is moderated by their satisfaction level.

While this dynamic mechanism explains the reciprocal relationship between stakeholders and a company's behavior, stakeholders' materiality is also contingent on a set of broader environmental and social drivers that are completely detached from a single company's control.

A clear example of this is the increased urgency linked to the stagnation and regression in achieving global SDGs targets (United Nations, 2023). Consider stakeholders' concerns regarding CO_2 emissions: the external pressure to reduce CO_2 may intensify over time as the effects of climate change become more evident and the sense of urgency increases, independently of any individual company's attention to the issue.

This book refers to the component of stakeholders' materiality that is independent from a single company's actions or behaviour using the term global materiality.

In conclusion, the sum of stakeholders' expectations on a sustainability issue—defined by aspects both dependent on and independent of a company's behaviour—represents a dynamic benchmark for determining what level of a company's performance can be considered sustainable in the context in which a certain organisation operates.

Expanding the class of stakeholders to include non-human actors (such as nature and the environment) in light of actor-network theory (Callon, 1984) significantly reinforces the meaning and representativity of this dynamic benchmark. Actor-network theory recognises non-human actors as active agents that can express their will through the intermediation of representatives, such as scientists, NGOs, or stakeholders in general. Viewed through this lens, stakeholders' pressure can be read as an attempt to give voice to nature (Quattrone, 2022). This perspective proposes a strong approach to sustainability (Wackernagel & Rees, 1996), directing external pressures to position societal and ecological issues at the centre of the company's system of values (Birkin, 1996), thereby shifting the focus of performance management activities to the sustainability indicators that truly matter (Daly, 1992).

Adopting this perspective effectively expands the described relationship with stakeholders into the broader field of impact management. This requires accounting for the total value added or subtracted to nature through business activities (Busco et al., 2024).

However, this comprehensive evaluation must not disregard the financial repercussions of sustainability performance. For this reason, Section 2.5 explains how being sustainable can also be considered a source of competitive advantage recognised by the market (Provaroni & Checcacci, 2024), capable of generating positive financial returns.

2.5 The Financial Dimension of Sustainability

Even though the ecological and social issues have assumed a global dimension and cannot any more be postponed, the sense of moral responsibility and the pressure exerted by stakeholders may not be sufficient to change the industrial purpose and align business activities with sustainability needs. In line with the *homo oeconomicus* classical economy perspective, financial measures still dominate the capitalistic system, shaping the decisions of individuals and collectives (Friedman, 1962). For this, the reasons behind the sustainability involvement of companies are often rooted in an economic—and not ecological—rationality (Banerjee, 2003).

This perspective calls for further analysis of the financial materiality approach, focusing on the sustainability repercussions on the entity's prospects (IFRS, 2023). Examples of financial repercussions can be found in the direct benefits of sustainability investments, such as the savings related to resource use efficiency (e.g., the reduction of energy or water consumption) or the mitigation of climate-related risks (e.g., investments that increase the resilience to adverse weather events).

Nevertheless, direct positive financial consequences can be appreciated only with reference to specific initiatives; more often, being sustainable means bearing direct costs which generate uncertain returns for the companies. This could be the case of investments in social sustainability, such as workers' well-being or gender equality, or environmental expenditures, such as the reduction of emissions or a more responsible supply chain management. In these cases, positive repercussions are not related to the single initiative but follow indirect paths such as the enhancement of the overall image of an organisation.

This is because the higher-scale financial benefits of companies' commitment to sustainability overcome organisational boundaries to be released to the entire community (Ekins & Zenghelis, 2021). On the one hand, such a systemic perspective may create space for policy interventions, stimulating shared commitment for the common good and the transition towards more sustainable economic paradigms. On the other hand, this could also question the entire meaning of sustainability reporting, with companies being required to take into account facts and issues that have global reach, and are therefore beyond the company's control (Gray, 2010).

In this context, reconducting the sustainability discourse to a manageable business dimension means focusing on the tangible consequences of investments and performance. To do this, all the facts should be analysed by considering the relationships with organisations' stakeholders, that is, those impacted by organisations' activities or that may recognise their value. For example, companies' sustainability performance is expected to affect positively the value of their shares recognised by investors and the market (Lopez et al., 2022), the interest rate conditions applied by banks (e.g., sustainability-linked loans), the possibility of accessing public funding opportunities, or customers' willingness to purchase (Collins et al., 2007).

Borrowing the return on investments (ROI) index from the financial field, academics often refer to the concept of sustainability (or social) return on investments (SROI) to indicate the sum of the direct and indirect financial positive consequences of sustainability performance (Atz et al., 2021; Nicholls et al., 2012). In this context, corporate reputation indexes emerge as the most influential factors in determining the sustainability financial returns, underlying the importance of a comprehensive and time-continuous strategy for sustainability (Orlitzky et al., 2003).

In conclusion and to summarise the discussion, the sustainability-related market is growing fast, even though the public attention is more related to the level of materiality of a theme on a global scale (i.e., global materiality) and to the SDGs-related challenges that nations must address, rather than to a single organisation's behaviour (United Nations Conference on Trade and Development [UNCTAD], 2023). However, the capacity of a company to satisfy the needs of its stakeholders is a factor that cannot be overlooked, since it represents the capacity of a company to meet the external expectations and to penetrate that expanding market.

For this reason, stakeholders' satisfaction is a crucial determinant of the financial repercussions of a company's sustainable performance (Orlitzky et al., 2003). Following this reasoning, Chap. 3 will present stakeholders' satisfaction as a central variable in the modelling process, used in this book as a proxy for determining the capacity of a company to achieve financial returns through sustainable performance.

Thanks to the adoption of an aggregated and systemic perspective, the literature review portrayed in this Chap. 2 has reconstructed the dimensions of corporate activities, both from the impact and financial perspectives. The materiality assessment, from which the analysis started, could also be seen as the conclusion of the analysis since the determinants of its impact and financial dimensions have been retraced and described.

This allows reading the sustainability reporting process as a closed system describing companies' dynamic answers to the changing nature of sustainability issues. For this reason, Systems Theories, and in particular System Dynamics, can support a more structured analysis of the sustainability reporting process, keeping the materiality analysis at the centre of the debate. Consequently, Chap. 3 presents the theoretical and methodological foundations of this book, followed by the description of the modelling process for creating a Sustainability Reporting Model.

References

Abhayawansa, S. (2022). Swimming against the tide: Back to single materiality for sustainability reporting. *Sustainability Accounting, Management and Policy Journal, 13*(6), 1361–1385. https://doi.org/10.1108/SAMPJ-07-2022-0378

Adams, C. A., & Abhayawansa, S. (2022). Connecting the COVID-19 pandemic, environmental, social and governance (ESG) investing and calls for 'harmonisation' of sustainability reporting. *Critical Perspectives on Accounting, 82*. https://doi.org/10.1016/j.cpa.2021.102309

Adams, C. A., Alhamood A., He X, Tian J., Wang, J., & Wang, L. (2021). *The double-materiality concept: Application and issues.*

Adams, C. A., Druckman, P. B., & Picot, R. C. (2020). *Sustainable development goals disclosure (SDGD) recommendations*. https://www.icas.com/professional-resources/sustainability/sustainable-development-goals/feedback-on-

Adams, C. A., & Mueller, F. (2022). Academics and policymakers at odds: The case of the IFRS foundation trustees' consultation paper on sustainability reporting. *Sustainability Accounting, Management and Policy Journal, 13*(6), 1310–1333. https://doi.org/10.1108/SAMPJ-10-2021-0436

Aji, M. H., & Sutikno, B. (2015). The extended consequence of greenwashing: Perceived consumer skepticism. *International Journal of Business and Information, 10*(4).

Albrecht, P., Burandt, S., & Schaltegger, S. (2007). Do sustainability projects stimulate organizational learning in universities? *International Journal of Sustainability in Higher Education, 8*(4), 403–415. https://doi.org/10.1108/14676370710823573

Al-Htaybat, K., & von Alberti-Alhtaybat, L. (2018). Integrated thinking leading to integrated reporting: Case study insights from a global player. *Accounting, Auditing and Accountability Journal, 31*(5), 1435–1460. https://doi.org/10.1108/AAAJ-08-2016-2680

AlWaer, H., Sibley, M., & Lewis, J. (2008). Different stakeholder perceptions of sustainability assessment. *Architectural Science Review, 51*(1), 48–59. https://doi.org/10.3763/asre.2008.5107

Arthur, W. B. (1989). Competing technologies, increasing returns, and lock-in by historical events. *The Economic Journal, 99*(394), 116. https://doi.org/10.2307/2234208

Arul, R., de Villiers, C., & Dimes, R. (2021). Insights from narrative disclosures regarding integrated thinking in integrated reports in South Africa and Japan. *Meditari Accountancy Research, 29*(4), 720–739. https://doi.org/10.1108/MEDAR-06-2020-0934

Atz, U., Van Holt, T., Douglas, E., & Whelan, T. (2021). The return on sustainability investment (ROSI): Monetizing financial benefits of sustainability actions in companies. In *Sustainable consumption and production, Volume II* (pp. 303–354). Springer International Publishing. https://doi.org/10.1007/978-3-030-55285-5_14

Banerjee, S. B. (2003). Who sustains whose development? Sustainable development and the reinvention of nature. *Organization Studies, 24*(1), 143–180. https://doi.org/10.1177/0170840603024001341

Bansal, P., & DesJardine, M. R. (2014). Business sustainability: It is about time. *Strategic Organization, 12*(1), 70–78. https://doi.org/10.1177/1476127013520265

Baret, P., & Helfrich, V. (2019). The "trilemma" of non-financial reporting and its pitfalls. *Journal of Management and Governance, 23*(2), 485–511. https://doi.org/10.1007/s10997-018-9430-z

Barnabè, F. (2008). Balanced scorecard e system dynamics. In C. Busco, A. Riccaboni, & A. Saviotti (Eds.), *Governance, Strategia e Misurazione delle Performance, Le nuove frontiere della Balanced Scorecard* (pp. 215–243). KNOWITÀ.

Berg, F., Kölbel, J. F., & Rigobon, R. (2022). Aggregate confusion: The divergence of ESG ratings. *Review of Finance, 26*(6), 1315–1344. https://doi.org/10.1093/rof/rfac033

Beske, F., Haustein, E., & Lorson, P. C. (2020). Materiality analysis in sustainability and integrated reports. *Sustainability Accounting, Management and Policy Journal, 11*(1), 162–186. https://doi.org/10.1108/SAMPJ-12-2018-0343

Bianchi, C. (2016). *Dynamic performance management* (Vol. 1). Springer International Publishing. https://doi.org/10.1007/978-3-319-31845-5

Birkin, F. (1996). The ecological accountant: From the cogito to thinking like a mountain. *Critical Perspectives on Accounting, 7*(3), 231–257. https://doi.org/10.1006/cpac.1996.0031

Bolt, R., & Tregidga, H. (2023). Methodological insights "materiality is …": Sensemaking and sensegiving through storytelling. *Accounting, Auditing and Accountability Journal, 36*(1), 403–427. https://doi.org/10.1108/AAAJ-06-2021-5314

Borgert, T., Donovan, J. D., Topple, C., & Masli, E. K. (2018). Initiating sustainability assessments: Insights from practice on a procedural perspective. *Environmental Impact Assessment Review, 72*, 99–107. https://doi.org/10.1016/j.eiar.2018.05.012

Busco, C., Consolandi, C., Malafronte, I., Sammarco, F., & Scognamiglio, E. (2024). *The impact of organizations: Measurement, Management and Corporate Reporting*. Giappichelli Editore, Routledge.

Busco, C., Granà, F., & Achilli, G. (2021). Understanding integrated thinking: Evidence from the field, the development of a framework and avenues for future research. *Meditari Accountancy Research, 29*(4), 673–690. https://doi.org/10.1108/MEDAR-04-2021-1263

Calabrese, A., Costa, R., Levialdi Ghiron, N., & Menichini, T. (2019). Materiality analysis in sustainability reporting: A tool for directing corporate sustainability towards emerging economic, environmental and social opportunities. *Technological and Economic Development of Economy, 25*(5), 1016–1038. https://doi.org/10.3846/tede.2019.10550

Calabrese, A., Costa, R., Levialdi, N., & Menichini, T. (2016). A fuzzy analytic hierarchy process method to support materiality assessment in sustainability reporting. *Journal of Cleaner Production, 121*, 248–264. https://doi.org/10.1016/j.jclepro.2015.12.005

Calabrese, A., Costa, R., & Rosati, F. (2015). A feedback-based model for CSR assessment and materiality analysis. *Accounting Forum, 39*(4), 312–327. https://doi.org/10.1016/j.accfor.2015.06.002

Callon, M. (1984). Some elements of a sociology of translation: Domestication of the scallops and the fishermen of St Brieuc Bay. *The Sociological Review, 32*(1_suppl), 196–233. https://doi.org/10.1111/j.1467-954X.1984.tb00113.x

Capelo, C., & Dias, J. F. (2009). A system dynamics-based simulation experiment for testing mental model and performance effects of using the balanced scorecard. *System Dynamics Review, 25*(1), 1–34. https://doi.org/10.1002/sdr.413

Carroll, A. B. (1991). The pyramid of corporate social responsibility: Toward the moral Management of Organizational Stakeholders. *Business Horizons, 34*, 4–39.

Carungu, J., Di Pietra, R., & Molinari, M. (2020). Mandatory vs voluntary exercise on non-financial reporting: Does a normative/coercive isomorphism facilitate an increase in quality? *Meditari Accountancy Research, 29*(3), 449–476. https://doi.org/10.1108/MEDAR-08-2019-0540

Catturi, G. (2021). *Principi di economia aziendale. L'azienda universale. L'idea forza, la morfologia e la fisiologia* (Settima Edizione). CEDAM.

Cerbone, D., & Maroun, W. (2020). Materiality in an integrated reporting setting: Insights using an institutional logics framework. *The British Accounting Review, 52*(3). https://doi.org/10.1016/j.bar.2019.100876

Cho, C. H., Laine, M., Roberts, R. W., & Rodrigue, M. (2015). Organized hypocrisy, organizational façades, and sustainability reporting. *Accounting, Organizations and Society, 40*, 78–94. https://doi.org/10.1016/j.aos.2014.12.003

Christensen, H. B., Hail, L., & Leuz, C. (2021). Mandatory CSR and sustainability reporting: Economic analysis and literature review. *Review of Accounting Studies, 26*(3), 1176–1248. https://doi.org/10.1007/s11142-021-09609-5

Clark, C. E. (2021). How do standard setters define materiality and why does it matter? *Business Ethics, the Environment & Responsibility, 30*(3), 378–391. https://doi.org/10.1111/beer.12351

Clarkson, P. M., Li, Y., Richardson, G. D., & Vasvari, F. P. (2008). Revisiting the relation between environmental performance and environmental disclosure: An empirical analysis. *Accounting, Organizations and Society, 33*(4–5), 303–327. https://doi.org/10.1016/j.aos.2007.05.003

Collins, C. M., Steg, L., & Koning, M. A. S. (2007). Customers' values, beliefs on sustainable corporate performance, and buying behavior. *Psychology and Marketing, 24*(6), 555–577. https://doi.org/10.1002/mar.20173

Connelly, B. L., Certo, S. T., Ireland, R. D., & Reutzel, C. R. (2011). Signaling theory: A review and assessment. *Journal of Management, 37*(1), 39–67. https://doi.org/10.1177/0149206310388419

Contrafatto, M. (2014). Stewardship theory: Approaches and perspectives. In *Accountability and social accounting for social and non-profit organizations* (pp. 177–196). https://doi.org/10.1108/S1041-706020140000017007

Cooper, S., & Michelon, G. (2022). Conceptions of materiality in sustainability reporting frameworks: Commonalities, differences and possibilities. In *Handbook of accounting and sustainability* (pp. 44–66). Edward Elgar Publishing. https://doi.org/10.4337/9781800373518.00010

Cosma, S., Leopizzi, R., Nobile, L., & Schwizer, P. (2022). Revising the non-financial reporting directive and the role of board of directors: A lost opportunity? *Journal of Applied Accounting Research, 23*(1), 207–226. https://doi.org/10.1108/JAAR-04-2021-0102

Crovini, C., Schaper, S., & Simoni, L. (2022). Dynamic accountability and the role of risk reporting during a global pandemic. *Accounting, Auditing and Accountability Journal, 35*(1), 169–185. https://doi.org/10.1108/AAAJ-08-2020-4793

Daly, H. E. (1992). Allocation, distribution, and scale: Towards an economics that is efficient, just, and sustainable. *Ecological Economics, 6*(3), 185–193. https://doi.org/10.1016/0921-8009(92)90024-M

Davis, J. H., Schoorman, F. D., & Donaldson, L. (1997). Toward a stewardship theory of management. *Academy of Management Review, 22*(1), 20–47. https://doi.org/10.5465/amr.1997.9707180258

de Freitas Netto, S. V., Sobral, M. F. F., Ribeiro, A. R. B., & Soares, G. R. d. L. (2020). Concepts and forms of greenwashing: A systematic review. *Environmental Sciences Europe, 32*(1). https://doi.org/10.1186/s12302-020-0300-3

Deegan, C. (2002). Introduction: The legitimising effect of social and environmental disclosures—A theoretical foundation. *Accounting, Auditing & Accountability Journal, 15*(3), 282–311. https://doi.org/10.1108/09513570210435852

Delmas, M. A., & Burbano, V. C. (2011). The drivers of greenwashing. *California Management Review, 54*(11).

De Micco, P., Rinaldi, L., Vitale, G., Cupertino, S., & Maraghini, M. P. (2020). The challenges of sustainability reporting and their management: The case of Estra. *Meditari Accountancy Research, 29*(3), 430–448. https://doi.org/10.1108/MEDAR-09-2019-0555

de Villiers, C., & Alexander, D. (2014). The institutionalisation of corporate social responsibility reporting. *The British Accounting Review, 46*(2), 198–212. https://doi.org/10.1016/j.bar.2014.03.001

de Villiers, C., La Torre, M., & Molinari, M. (2022). The global reporting initiative's (GRI) past, present and future: Critical reflections and a research agenda on sustainability reporting (standard-setting). *Pacific Accounting Review, 34*(5), 728–747. https://doi.org/10.1108/PAR-02-2022-0034

Di Vaio, A., Syriopoulos, T., Alvino, F., & Palladino, R. (2020). "Integrated thinking and reporting" towards sustainable business models: A concise bibliometric analysis. *Meditari Accountancy Research, 29*(4), 691–719). Emerald Group Holdings Ltd. https://doi.org/10.1108/MEDAR-12-2019-0641

Dimes, R., & de Villiers, C. (2020). How management control systems enable and constrain integrated thinking. *Meditari Accountancy Research, 29*(4), 851–872. https://doi.org/10.1108/MEDAR-05-2020-0880

Diouf, D., & Boiral, O. (2017). The quality of sustainability reports and impression management: A stakeholder perspective. *Accounting, Auditing and Accountability Journal, 30*(3), 643–667. https://doi.org/10.1108/AAAJ-04-2015-2044

Eccles, R. G., Ioannou, I., & Serafeim, G. (2014). The impact of corporate sustainability on organizational processes and performance. *Management Science, 60*(11), 2835–2857. https://doi.org/10.1287/mnsc.2014.1984

Eccles, R. G., Lee, L.-E., & Stroehle, J. C. (2020). The social origins of ESG: An analysis of Innovest and KLD. *Organization & Environment, 33*(4), 575–596. https://doi.org/10.1177/1086026619888994

Eccles, R. G., & Youmans, T. (2016). Materiality in corporate governance: The statement of significant audiences and materiality. *Journal of Applied Corporate Finance, 28*(2), 39–46. https://doi.org/10.1111/jacf.12173

Edgley, C. (2014). A genealogy of accounting materiality. *Critical Perspectives on Accounting, 25*(3), 255–271. https://doi.org/10.1016/j.cpa.2013.06.001

Edgley, C., Jones, M. J., & Atkins, J. (2015). The adoption of the materiality concept in social and environmental reporting assurance: A field study approach. *British Accounting Review, 47*(1), 1–18. https://doi.org/10.1016/j.bar.2014.11.001

EFRAG. (2021). *Proposals for a relevant and dynamic EU sustainability reporting standard-setting*.

Eilifsen, A., & Messier, W. F. (2015). Materiality guidance of the major public accounting firms. *Auditing: A Journal of Practice & Theory, 34*(2), 3–26. https://doi.org/10.2308/ajpt-50882

Ekins, P., & Zenghelis, D. (2021). The costs and benefits of environmental sustainability. *Sustainability Science, 16*(3), 949–965. https://doi.org/10.1007/s11625-021-00910-5

Elkington, J. (1994). Towards the sustainable corporation: Win-win-win business strategies for sustainable development. *California Management Review, 36*(2), 90–100. https://doi.org/10.2307/41165746

European Commission (2023). *Commission Delegated Regulation (EU) 2023/2772 of 31 July 2023 supplementing directive 2013/34/EU of the European Parliament and of the council as regards sustainability reporting standards.*

Farooq, M. B., & de Villiers, C. (2019). Understanding how managers institutionalise sustainability reporting: Evidence from Australia and New Zealand. *Accounting, Auditing and Accountability Journal, 32*(5), 1240–1269. https://doi.org/10.1108/AAAJ-06-2017-2958

Fasan, M., & Mio, C. (2017). Fostering stakeholder engagement: The role of materiality disclosure in integrated reporting. *Business Strategy and the Environment, 26*(3), 288–305. https://doi.org/10.1002/bse.1917

Fernandez-Feijoo, B., Romero, S., & Ruiz, S. (2014). Effect of stakeholders' pressure on transparency of sustainability reports within the GRI framework. *Journal of Business Ethics, 122*(1), 53–63. https://doi.org/10.1007/s10551-013-1748-5

Fernando, S., & Lawrence, S. (2014). A theoretical framework for CSR practices: Integrating legitimacy theory, stakeholder theory and institutional theory. *Journal of Theoretical Accounting Research, 10*(1), 149–178.

Ferrero-Ferrero, I., Fernández-Izquierdo, M. Á., Muñoz-Torres, M. J., & Bellés-Colomer, L. (2018). Stakeholder engagement in sustainability reporting in higher education. *International Journal of Sustainability in Higher Education, 19*(2), 313–336. https://doi.org/10.1108/IJSHE-06-2016-0116

Ferrero-Ferrero, I., León, R., & Muñoz-Torres, M. J. (2021). Sustainability materiality matrices in doubt: May prioritizations of aspects overestimate environmental performance? *Journal of Environmental Planning and Management, 64*(3), 432–463. https://doi.org/10.1080/09640568.2020.1766427

Figge, F., Hahn, T., Schaltegger, S., & Wagner, M. (2002). The sustainability balanced scorecard—Linking sustainability management to business strategy. *Business Strategy and the Environment, 11*(5), 269–284. https://doi.org/10.1002/bse.339

Font, X., Guix, M., & Bonilla-Priego, M. J. (2016). Corporate social responsibility in cruising: Using materiality analysis to create shared value. *Tourism Management, 53*, 175–186. https://doi.org/10.1016/j.tourman.2015.10.007

Forrester, J. W. (1961). *Industrial dynamics*. Massachusetts Institute of Technology (MIT).

Fortanier, F., Kolk, A., & Pinkse, J. (2011). Harmonization in CSR reporting: MNEs and global CSR standards. *Management International Review, 51*(5), 665–696. https://doi.org/10.1007/s11575-011-0089-9

Frankel, M. (1962). The production function in allocation and growth: A synthesis. *The American Economic Review, 52*(5), 996–1022.

Freeman, R. E. (1984). *Strategic management. A stakeholder approach*. Pitman.

Freeman, R. E. (2008). Ending the so-called "Friedman-Freeman" debate. *Business Ethics Quarterly, 18*(2), 162–166.

Freiberg, D., Rogers, J., & Serafeim, G. (2019). *How ESG issues become financially material to corporations and their investors*.

Friedman, M. (1962). *Capitalism and freedom*. University of Chicago Books.

Garst, J., Maas, K., & Suijs, J. (2022). Materiality assessment is an art, not a science: Selecting ESG topics for sustainability reports. *California Management Review, 65*(1), 64–90. https://doi.org/10.1177/00081256221120692

Gerwanski, J., Kordsachia, O., & Velte, P. (2019). Determinants of materiality disclosure quality in integrated reporting: Empirical evidence from an international setting. *Business Strategy and the Environment, 28*(5), 750–770. https://doi.org/10.1002/bse.2278

Gray, R. (2010). Is accounting for sustainability actually accounting for sustainability…and how would we know? An exploration of narratives of organisations and the planet. *Accounting, Organizations and Society, 35*(1), 47–62. https://doi.org/10.1016/j.aos.2009.04.006

GRI. (2002). *G2 Sustainability Reporting Guidelines 2002*.

GRI. (2013). *G4 guidelines*.

GRI. (2022). *GRI 3: Material Topics 2021*.

Guix, M., & Font, X. (2020). The materiality balanced scorecard: A framework for stakeholder-led integration of sustainable hospitality management and reporting. *International Journal of Hospitality Management, 91*. https://doi.org/10.1016/j.ijhm.2020.102634

Guthrie, J., Manes-Rossi, F., & Orelli, R. L. (2017). Integrated reporting and integrated thinking in Italian public sector organizations. *Meditari Accountancy Research, 25*(4), 553–573. https://doi.org/10.1108/MEDAR-06-2017-0155

Guthrie, J., & Parker, L. D. (1989). Corporate social reporting: A rebuttal of legitimacy theory. *Accounting and Business Research, 19*(76), 343–352. https://doi.org/10.1080/00014788.1989.9728863

Hahn, R., & Kühnen, M. (2013). Determinants of sustainability reporting: A review of results, trends, theory, and opportunities in an expanding field of research. *Journal of Cleaner Production, 59*, 5–21. https://doi.org/10.1016/j.jclepro.2013.07.005

Hansen, E. G., & Schaltegger, S. (2016). The sustainability balanced scorecard: A systematic review of architectures. *Journal of Business Ethics, 133*(2), 193–221. https://doi.org/10.1007/s10551-014-2340-3

Herath, R., Senaratne, S., & Gunarathne, N. (2021). Integrated thinking, orchestration of the six capitals and value creation. *Meditari Accountancy Research, 29*(4), 873–907. https://doi.org/10.1108/MEDAR-01-2020-0676

Hicks, E. L. (1964). Materiality. *Journal of Accounting Research, 2*(2), 158–171.

Hoffman, A. J., & Woody, J. G. (2008). *Climate change: What's your business strategy?* Harvard Business Press.

IFRS. (2023). *IFRS S1 general requirements for disclosure of sustainability-related financial information*.

IIRC. (2021). *International <IR> framework*. www.integratedreporting.org

Jones, E. (2019). Rethinking greenwashing: Corporate discourse, unethical practice, and the unmet potential of ethical consumerism. *Sociological Perspectives, 62*(5), 728–754. https://doi.org/10.1177/0731121419849095

Jørgensen, S., Mjøs, A., & Pedersen, L. J. T. (2022). Sustainability reporting and approaches to materiality: Tensions and potential resolutions. *Sustainability Accounting, Management and Policy Journal, 13*(2), 341–361. https://doi.org/10.1108/SAMPJ-01-2021-0009

Kaplan, R. S., & Norton, D. P. (1992). The balanced scorecard: Measures that drive performance. *Harvard Business Review., 70*, 71.

Kaplan, R. S., & Norton, D. P. (1996). *Using the balanced scorecard as a strategic management system*. Harvard Business Review.

Kaplan, R. S., & Norton, D. P. (2000). *Having trouble with your strategy? Then map it*. Harvard Business Review.

Kaplan, R. S., & Norton, D. P. (2004). *Strategy maps. Converting intangible assets into tangible outcomes*. Harvard Business School Press.

Khan, M., Serafeim, G., & Yoon, A. (2016). Corporate sustainability: First evidence on materiality. *The Accounting Review, 91*(6), 1697–1724. https://doi.org/10.2308/accr-51383

Kuh, T., Shepley, A., Bala, G., & Flowers, Mi. (2020). *Dynamic materiality: Measuring what matters*. https://ssrn.com/abstract=3521035.

Kunc, M., Giorgino, M. C., & Barnabè, F. (2021). Developing forward-looking orientation in integrated reporting. *Meditari Accountancy Research, 29*(4), 823–850. https://doi.org/10.1108/MEDAR-12-2019-0664

La Torre, M., Sabelfeld, S., Blomkvist, M., & Dumay, J. (2020). Rebuilding trust: Sustainability and non-financial reporting and the European Union regulation. *Meditari Accountancy Research, 28*(5), 701–725. https://doi.org/10.1108/MEDAR-06-2020-0914

Lai, A., Melloni, G., & Stacchezzini, R. (2017). What does materiality mean to integrated reporting preparers? An empirical exploration. *Meditari Accountancy Research, 25*(4), 533–552. https://doi.org/10.1108/MEDAR-02-2017-0113

Larrinaga, C. (2007). Sustainability reporting: Insights from neo-institutional theory. In J. Unerman & J. Bebbington (Eds.), *Sustainability accounting and accountability*. Routledge.

Lazarus, R. S., & Folkman, S. (1984). *Stress, appraisal, and coping*. Springer Publishing Company.

Leavy, B. (2012). Getting back to what matters—Creating long-term economic and social value. *Strategy & Leadership, 40*(4), 12–20. https://doi.org/10.1108/10878571211242902

Locke, E. A., & Latham, G. P. (1990). *A theory of goal setting and task performance*. Prentice-Hall.

Lopez, B., Rangel, C., & Fernández, M. (2022). The impact of corporate social responsibility strategy on the management and governance axis for sustainable growth. *Journal of Business Research, 150*, 690–698. https://doi.org/10.1016/j.jbusres.2022.06.025

Machado, B. A. A., Dias, L. C. P., & Fonseca, A. (2021). Transparency of materiality analysis in GRI-based sustainability reports. *Corporate Social Responsibility and Environmental Management, 28*(2), 570–580. https://doi.org/10.1002/csr.2066

Macias, H. A., & Farfan-Lievano, A. (2017). Integrated reporting as a strategy for firm growth: Multiple case study in Colombia. *Meditari Accountancy Research, 25*(4), 605–628. https://doi.org/10.1108/MEDAR-11-2016-0099

Manetti, G. (2011). The quality of stakeholder engagement in sustainability reporting: Empirical evidence and critical points. *Corporate Social Responsibility and Environmental Management, 18*(2), 110–122. https://doi.org/10.1002/csr.255

Maniora, J. (2018). Mismanagement of sustainability: What business strategy makes the difference? Empirical evidence from the USA. *Journal of Business Ethics, 152*(4), 931–947. https://doi.org/10.1007/s10551-018-3819-0

Meadows, D. H. (2008). *Thinking in systems* (Sustainability institute, Ed.).
Melloni, G., Stacchezzini, R., & Lai, A. (2016). The tone of business model disclosure: An impression management analysis of the integrated reports. *Journal of Management and Governance, 20*(2), 295–320. https://doi.org/10.1007/s10997-015-9319-z
Merchant, K. A. (1982). The control function of management. *Sloan Management Review, 24*(4).
Merchant, K. A., & Van der Stede, W. A. (2007). *Management control systems: Performance measurement, evaluation and incentives*. Pearson Education.
Meyer, J. W., & Rowan, B. (1977). Institutionalized organizations: Formal structure as myth and ceremony. *American Journal of Sociology, 83*(2), 340–363. https://doi.org/10.1086/226550
Millar, E., & Searcy, C. (2020). The presence of citizen science in sustainability reporting. *Sustainability Accounting, Management and Policy Journal, 11*(1), 31–64. https://doi.org/10.1108/SAMPJ-01-2019-0006
Molinari, M., Maraghini, M. P., & Riccaboni, A. (2021). Reporting di Sostenibilità e Controllo Manageriale: L'esperienza di Edison S.p. A. *Management Control, 2*, 61–86. https://doi.org/10.3280/maco2021-002004
Morecroft, J. D. W. (2015). *Strategic modelling and business dynamics. A feedback systems approach* (2nd ed.). Wiley.
Moynihan, D. (2008). *The dynamics of performance management*. Georgetown University Press.
Nicholls, J., Lawlor, E., Neitzert, E., & Goodspeed, T. (2012). *A guide to social return on investment*. The SROI Network.
O'Donovan, G. (2002). Environmental disclosures in the annual report. *Accounting, Auditing & Accountability Journal, 15*(3), 344–371. https://doi.org/10.1108/09513570210435870
OIBR. (2022). *Operationalising materiality–applied guidelines on how to identify and monitor the evolution of sustainability-related material issues.*
Oliver, R. L. (1980). A cognitive model of the antecedents and consequences of satisfaction decisions. *Journal of Marketing Research, 17*(4), 460–469. https://doi.org/10.1177/002224378001700405
Orlitzky, M., Schmidt, F. L., & Rynes, S. L. (2003). Corporate social and financial performance: A meta-analysis. *Organization Studies, 24*(3), 403–441. www.sagepublications.com
Pazienza, M., de Jong, M., & Schoenmaker, D. (2023). Why corporate sustainability is not yet measured. *Sustainability, 15*(7), 6275. https://doi.org/10.3390/su15076275
Porter, M. E., & Kramer, M. R. (2006). *Strategy and society: The link between competitive advantage and corporate social responsibility*. Harvard Business Review.
Prado-Lorenzo, J. M., Gallego-Alvarez, I., & Garcia-Sanchez, I. M. (2009). Stakeholder engagement and corporate social responsibility reporting: The ownership structure effect. *Corporate Social Responsibility and Environmental Management, 16*(2), 94–107. https://doi.org/10.1002/csr.189
Provaroni, L., & Checcacci, F. (2024). The impact of corporate social responsibility on financial performance and firm value. In C. Busco, C. Consolandi, I. Malafronte, F. Sammarco, & E. Scognamiglio (Eds.), *The impact of organizations: Measurement, management and corporate reporting* (pp. 15–30). Giappichelli Editore, Routledge.
Puroila, J., & Mäkelä, H. (2019). Matter of opinion: Exploring the socio-political nature of materiality disclosures in sustainability reporting. *Accounting, Auditing and Accountability Journal, 32*(4), 1043–1072. https://doi.org/10.1108/AAAJ-11-2016-2788
Quarchioni, S., Ruggiero, P., & Damiano, R. (2021). Flows of information and meaning: A vocabulary approach to integrated thinking and reporting. *Meditari Accountancy Research, 29*(4), 740–774. https://doi.org/10.1108/MEDAR-01-2020-0677
Quattrone, P. (2022). Seeking transparency makes one blind: How to rethink disclosure, account for nature and make corporations sustainable. *Accounting, Auditing and Accountability Journal, 35*(2), 547–566. https://doi.org/10.1108/AAAJ-04-2021-5233
Remmer, S., & Gilbert, D. U. (2019). Applying materiality assessment in strategic management: The implicit coating of the materiality lens. In T. Wunder (Ed.), *Rethinking strategic management. CSR, sustainability, ethics & governance* (pp. 267–291). Springer. https://doi.org/10.1007/978-3-030-06014-5_13

Rinaldi, L., Unerman, J., & de Villiers, C. (2018). Evaluating the integrated reporting journey: Insights, gaps and agendas for future research. *Accounting, Auditing and Accountability Journal, 31*(5), 1294–1318. https://doi.org/10.1108/AAAJ-04-2018-3446

Rockström, J., Steffen, W., Noone, K., Persson, Å., Chapin, F. S., Lambin, E. F., Lenton, T. M., Scheffer, M., Folke, C., Schellnhuber, H. J., Nykvist, B., de Wit, C. A., Hughes, T., van der Leeuw, S., Rodhe, H., Sörlin, S., Snyder, P. K., Costanza, R., Svedin, U., et al. (2009). A safe operating space for humanity. *Nature, 461*(7263), 472–475. https://doi.org/10.1038/461472a

Schiehll, E., & Kolahgar, S. (2021). Financial materiality in the informativeness of sustainability reporting. *Business Strategy and the Environment, 30*(2), 840–855. https://doi.org/10.1002/bse.2657

Spaargaren, G., & Mol, A. P. J. (1992). Sociology, environment, and modernity: Ecological modernization as a theory of social change. *Society & Natural Resources, 5*(4), 323–344. https://doi.org/10.1080/08941929209380797

Sterman, J. D. (2000). *Business dynamics: Systems thinking and modeling for a complex world.* Irwin McGraw-Hill.

Sundin, H., Granlund, M., & Brown, D. A. (2010). Balancing multiple competing objectives with a balanced scorecard. *European Accounting Review, 19*(2), 203–246. https://doi.org/10.1080/09638180903118736

Teece, D. J., Pisano, G., & Shuen, A. (1997). Dynamic capabilities and strategic management. *Strategic Management Journal, 18*, 509–533.

Tirado-Valencia, P., Cordobés-Madueño, M., Ruiz-Lozano, M., & De Vicente-Lama, M. (2021). Integrated thinking in the integrated reports of public sector companies. Evidence and contextual factors. *Sustainability Accounting, Management and Policy Journal, 12*(2), 330–352. https://doi.org/10.1108/SAMPJ-11-2019-0387

Torelli, R., Balluchi, F., & Furlotti, K. (2020a). The materiality assessment and stakeholder engagement: A content analysis of sustainability reports. *Corporate Social Responsibility and Environmental Management, 27*(2), 470–484. https://doi.org/10.1002/csr.1813

Torelli, R., Balluchi, F., & Lazzini, A. (2020b). Greenwashing and environmental communication: Effects on stakeholders' perceptions. *Business Strategy and the Environment, 29*(2), 407–421. https://doi.org/10.1002/bse.2373

Tregidga, H., Milne, M. J., & Kearins, K. (2018). Ramping up resistance: Corporate sustainable development and academic research. *Business & Society, 57*(2), 292–334. https://doi.org/10.1177/0007650315611459

Tsoukas, H., & Chia, R. (2002). On organizational becoming: Rethinking organizational change. *Organization Science, 13*(5), 567–582. https://doi.org/10.1287/orsc.13.5.567.7810

UNCTAD–United Nations Conference on Trade and Development. (2023). *World Investment Report 2023: Investment and sustainable energy.*

Unerman, J., Bebbington, J., & O'Dwyer, B. (2018). Corporate reporting and accounting for externalities. *Accounting and Business Research, 48*(5), 497–522. https://doi.org/10.1080/00014788.2018.1470155

United Nations. (2023). *Progress towards the sustainable development goals: Towards a rescue plan for people and planet report of the secretary-general (Special Edition).*

Uyar, A., Karaman, A. S., & Kilic, M. (2020). Is corporate social responsibility reporting a tool of signaling or greenwashing? Evidence from the worldwide logistics sector. *Journal of Cleaner Production, 253*, 119997. https://doi.org/10.1016/j.jclepro.2020.119997

Vignieri, V. (2022). Enhancing performance regimes to enable outcome-based policy analysis in cross-boundary settings (Vol. 6). Springer International Publishing. doi:https://doi.org/10.1007/978-3-031-07074-7.

Wackernagel, M., & Rees, W. (1996). *Our ecological footprint: Reducing human impact on the earth.* New Society Publishers.

Weber, H., & Weber, M. (2020). When means of implementation meet ecological modernization theory: A critical frame for thinking about the sustainable development goals initiative. *World Development, 136*, 105129. https://doi.org/10.1016/j.worlddev.2020.105129

WCED, World Commission on Environment and Development (1987), *Our common future*. Available at: https://doi.org/un-documents.net/our-common-future.pdf

World Economic Forum. (2020). Embracing the new age of materiality. *Harnessing the Pace of Change in ESG*. www.weforum.org

Yusliza, M. Y., Yong, J. Y., Tanveer, M. I., Ramayah, T., Noor Faezah, J., & Muhammad, Z. (2020). A structural model of the impact of green intellectual capital on sustainable performance. *Journal of Cleaner Production, 249*. https://doi.org/10.1016/j.jclepro.2019.119334

Zarzycka, E., & Krasodomska, J. (2021). Environmental key performance indicators: The role of regulations and stakeholder influence. *Environment Systems and Decisions, 41*(4), 651–666. https://doi.org/10.1007/s10669-021-09825-z

Zharfpeykan, R. (2021). Representative account or greenwashing? Voluntary sustainability reports in Australia's mining/metals and financial services industries. *Business Strategy and the Environment, 30*(4), 2209–2223. https://doi.org/10.1002/bse.2744

Zhou, Y. (2011). *Materiality approach in sustainability reporting: Applications, dilemmas, and challenges* (pp. 1–12). 1st World Sustainability Forum.

Chapter 3
The "Sustainability Reporting Model"

Contents

3.1 System Dynamics Soft and Generic Modelling Process

3.1.1 The Need for a System Dynamics Analysis of the Sustainability Reporting Process

Building on the systemic view of sustainability reporting developed in Chap. 2, this chapter introduces System Dynamics as a methodological framework for modelling the long-term consequences of corporate sustainability strategies. The sustainability

R. Santoni, *The Dynamics of Materiality*, System Dynamics for Performance Management & Governance 8, https://doi.org/10.1007/978-3-032-19280-6_3

reporting process is characterised by complex interactions between internal priorities and external expectations, often exacerbated by inaccurate materiality assessments and the strategic use of misleading communication techniques such as greenwashing.

These dynamics unfold over time and are influenced by feedback mechanisms, delays in stakeholder reactions, and the evolving relevance of sustainability issues in the public discourse. For instance, the loss of stakeholder trust following perceived greenwashing, or the time required to integrate materiality into strategic planning, can significantly affect corporate performance and legitimacy.

To address this complexity, System Dynamics offers a set of tools capable of representing causal relationships, feedback loops, and accumulation processes. The following sections present the theoretical foundations of the method and outline the modelling approach adopted in this book, which aims to capture the systemic nature of sustainability reporting and support strategic decision-making through simulation.

3.1.2 System Dynamics Roots and Building Blocks

Several theories and methods can support the development of the systemic perspective introduced in Sect. 3.1.1 (see Cabrera et al., 2008). Among these, System Dynamics, first proposed by Forrester (1961), stands out for its effectiveness in analysing recurring and chronic problems, such as persistent tensions like greenwashing activities or the protracted low corporate commitment to sustainability. This section presents System Dynamics both as a structural theory and a practical method, highlighting its suitability for modelling social systems and offering tools to represent structures related to sustainability reporting and simulate future scenarios with multiple actors involved (Größler et al., 2008).

System Dynamics is part of a broader transdisciplinary tradition rooted in general Systems Theories, which posit that organised wholes of any kind can be described using a shared formal apparatus, thanks to the identification of common and invariant structures and mechanisms (Boulding, 1956; Von Bertalanffy, 1968). A system can be understood as a set of interdependent variables whose interactions—both among internal components and with the external environment—generate complex dynamics. Within this framework, one of the distinctive features of System Dynamics is its problem orientation: The goal is not to replicate system behaviour per se, but to identify the systemic roots of a problem in order to support its resolution and management (Barlas, 2002).

To achieve this managerial purpose, System Dynamics draws on cybernetics and servomechanisms to design feedback and control structures within dynamic systems, supported by information technologies and simulation methods (Brown & Campbell, 1948; Schwaninger, 2009; Wiener, 1948).

Originally developed at MIT with the publication of "Industrial Dynamics" by Jay Forrester (1961), System Dynamics explores the principal interdependencies within social systems by constructing simulation models that incorporate feedback loops and accumulation processes, key elements that define system behaviour over time (Größler et al., 2008).

Its application in business studies is closely linked to the systemic approach to business administration, which views companies as systems of economic forces or operations that coordinate and integrate internal and external factors to pursue long-term economic equilibrium (Amaduzzi, 1953; Forrester & Amaduzzi, 1974; Giannessi, 1960; Onida, 1953). This approach has been further developed in managerial studies, where clarifying causal linkages among system components is considered essential for making mental models explicit and understanding the means-end relationships that underpin performance management (Bianchi, 2016; Oladimeji et al., 2020).

A critical aspect of System Dynamics–based studies and applications is the learning purpose and orientation (Vignieri, 2022): the method is not designed to produce precise forecasts, but rather to foster understanding of the emerging dynamics that shape possible scenarios, helping to identify leverage points and implement policies that guide system behaviour towards desirable outcomes (D. H. Meadows, 2008).

Schwaninger (2009, pp. 12–13) identifies the following as key features of System Dynamics:

1. A focus on feedback loops and the dynamics they generate within networks of interrelated variables
2. A high degree of operational clarity, with formal modelling based on the mathematical quantification of graphically mapped relationships
3. The integration of both qualitative and quantitative aspects of modelling and simulation, aiming at gaining insights into system behaviour rather than producing exact predictions
4. Broad applicability across disciplines
5. Availability of advanced software tools for implementation
6. Potential synergies with other tools and methodologies

The system object of study can be identified in a complex collection of interrelated elements, organised to form a meaningful whole aimed at pursuing a specific purpose (Meadows, 2008). Although System Dynamics is a discipline devoted to the study of systems, it does not seek to replicate overly complicated or poorly understood systems in their entirety. Instead, it relies on simplified, symbolic, dynamic, descriptive, and time-continuous representations, called models, designed to capture the essential structures and behaviours of the system. These models are constructed by selecting meaningful aspects of system architectures in order to uncover the causes of specific practical or theoretical problems (Barlas, 2009, p. 1134).

As previously discussed, problem orientation is a crucial component of the System Dynamics approach. It guides the identification of critical variables that must be included within the model boundaries to effectively describe the system's future development trajectories (Sterman, 2000, Chap. 4).

This orientation is essential for transitioning from the theoretical foundations of System Dynamics to its operational application. Indeed, the applicative part consists in explaining the determinants of the dynamic problems under investigation through three core concepts: (1) causality; (2) feedback loops; and (3) accumulation processes and delays.

Starting with causality, a system's structure is composed of causal relationships among its components. Causation refers to the influence that one variable exerts on another. The polarity of a causal relationship is positive when a change in the independent variable causes a change in the same direction in the dependent variable—typically represented by a plus sign (+) and a blue arrow. Conversely, the polarity is negative when the change occurs in the opposite direction, indicated by a minus sign (−) and a red arrow. While one-way causality is static, feedback relationships lie at the heart of System Dynamics and must be explicitly described (Forrester, 1961).

A variable is part of a feedback loop when a change in its value eventually influences itself, often with a delay (Größler et al., 2008). A feedback loop consists of a closed chain of relationships that passes through a set of decisions, rules, and mechanisms by which the system regulates itself (Meadows, 2008, p. 51). There are two main types of feedback loops: reinforcing (positive), denoted by "R", and balancing (negative or goal-seeking), denoted by "B". A classic example of a reinforcing loop is the relationship between population and births: a larger population leads to more births, which in turn further increases the population. In contrast, a balancing loop might involve population and deaths: A larger population results in higher mortality, which acts to stabilise the population size. However, forecasting population dynamics is not straightforward, as causes and effects are often separated in space and time.

This complexity arises from the presence of accumulations, such as population, which generate delayed consequences, like the effects of overcrowding on food availability or birth and death rates. System Dynamics captures these characteristics by incorporating time delays and accumulation processes into its models. Delays can affect both material and informational flows, occurring when a change in one variable influences another only after a certain period (Sterman, 2000).

The idea of accumulation is at the core of two building blocks of dynamic systems, namely the stocks and the flows. Stocks—typically represented by rectangles—accumulate values over time. The processes of accumulation and depletion require time and introduce inertia and memory into the system, slowing down its response to change (Meadows, 2008). Stocks create delays by accumulating (or releasing) the difference between their rates of change, known as flows. Flows are depicted as pipes controlled by valves, pointing into (inflows) or out of (outflows) the stock (Sterman, 2000).

While problem orientation, feedback causality, and accumulation/delay mechanisms are foundational to all System Dynamics models, additional considerations are needed before presenting the Sustainability Reporting Model. This is due to the inherently subjective nature of social systems, which are composed of intangible and soft components. These elements require specific justification before they can be translated into quantitative formulations, a topic addressed in the following Subsection 3.1.3.

3.1.3 Soft and Generic Quantitative Modelling

Although System Dynamics relies on a shared language built around the foundational elements presented in Sect. 3.1.2, alternative approaches to the discipline exist. In some interpretations, System Dynamics is viewed as a positivist method, applicable only when relationships among variables can be objectively represented through mathematical formulations derived from natural laws or empirical studies. In contrast, other scholars and practitioners argue that the benefits of the modelling process are so significant that its application should not be restricted to systems falling strictly within the realm of positivism. In this view, the emphasis shifts from the precision of individual relationships to the dominance of the overall system structure in shaping behaviour. This has led to the development of modelling techniques capable of addressing all types of challenges through qualitative and quantitative causal representations (see Coyle, 2000; Richardson, 1997).

As discussed in Chap. 2, the landscape of sustainability reporting is characterised by intangible dimensions, such as the different interpretations of materiality, which are often rooted in the underlying meaning attributed to the reporting process itself (see Sect. 2.3). For this reason, this book adopts an interpretivist approach to modelling and provides, in this section, the methodological rationale supporting that choice.

A key aspect concerns the construction of a quantitative model based on soft variables, such as materiality, expectations, satisfaction, and scepticism. Several arguments support the use of quantitative modelling even in immaterial contexts. These arguments begin with a recognition of the limitations of purely qualitative modelling tools, such as causal loop diagrams (CLDs). CLDs are graphical tools valuable for mapping complex interactions among multiple variables, identifying the principal feedback loops and thus supporting managerial decision-making (Morecroft, 1985; Wolstenholme & Coyle, 1983). However, despite the causal and feedback relationship identification, CLDs cannot fully capture the dynamic complexity of systems, particularly in terms of how the dynamic behaviour unfolds over time. This is because without quantitative data, it becomes impossible to assess whether a variable will tend to increase or decrease, or whether shifts in loop dominance might occur.

Moreover, concepts such as accumulation and delay, introduced in Sect. 3.1.2, can only be represented graphically in qualitative models. For instance, stock and flow diagrams (SFDs) enhance CLDs by indicating where stocks and accumulation processes are located, but they still lack the ability to convey the timing and magnitude of changes. In contrast, quantitative System Dynamics enables mathematical simulation, allowing for the exploration of complex causal maps and making the method suitable for addressing issues involving soft variables and incomplete data (Homer & Oliva, 2001; Richardson, 1999), as in the case of this book.

Before moving on, it is important to clarify that "soft" refers to the qualitative nature of a variable, not to its uncertainty or lack of measurement. Soft variables may be difficult to estimate or standardise, but they are essential to understanding system behaviour (Mclucas, 2003). Indeed, omitting these variables from model formulation means assuming they have no impact, a modelling assumption that—as Forrester (1961, p. 57) noted—is usually incorrect. Therefore, quantitative soft modelling does not aim to produce precise forecasts, but rather to preserve the benefits of simulation in studying trends and long-term consequences of alternative scenarios and managerial decisions (Richardson, 1986; Schwaninger, 2009).

As concerns the development of this book, different modelling options were considered. These mainly refer to the development of a model based on one or more case studies, using empirical data to quantify relationships. While valid, this approach was not aligned with the book's objective, which is to identify the performance drivers that shape firms' sustainability transitions over time. Anchoring the model to a specific company's past experience would have limited its meaning, potentially reducing it to a best-in-class example or narrowing its relevance to a particular business–stakeholder dynamic.

Instead, the approach adopted here draws inspiration from the work of Winch et al. (1999), which introduced the concept of generic model structures to support managers to govern change in business dynamics. System Dynamics simulators based on generic models aim to challenge managers' pre-existing mental models in contexts where future conditions may differ significantly from current ones (see also Winch & Arthur, 2002). These ideas are particularly relevant to sustainability transitions, where business-as-usual practices are increasingly challenged by climate change and other systemic pressures.

In line with Winch et al. (1999, pp. 1–2), the rationale for adopting a generic model structure in this book is threefold:

- To enable learning transference across industry sectors, by focusing on topics relevant to all organisations, such as materiality and sustainability performance
- To prioritise general relationships over particularities, given the lack of standardisation in reporting guidelines and the need to understand systemic dynamics
- To develop a generally representative framework that can be parameterised for specific cases, serving as a starting point for future research and practical applications

With these premises in place, the following Subsection 3.1.4 presents the modelling framework used to develop the generic structure. It outlines the process of model formulation and validation, which aims to challenge managerial mental models and support strategic thinking in the face of sustainability-related challenges and opportunities.

3.1.4 The Modelling Process and Phases

The steps of the modelling process presented here are based on the widely recognised framework proposed by Sterman (2000, Chap. 3). This scheme is adopted for its clarity and broad acceptance, although it represents just one of several possible approaches to modelling (Randers, 1997; Richardson & Pugh, 1997). While the steps are presented in a linear sequence, the modeller is engaged in a continuous learning process of revision, questioning, testing, and refinement, where each phase is closely connected to the others in an iterative approach (Homer, 1996).

The first step, developed in Sections 3.2.1 and 3.2.2, is problem articulation and boundary selection. This involves identifying, selecting, and describing the main variables that constitute the problem, its underlying causes, and the time horizon of analysis. The starting point of the simulation should include the roots of the problem, while the endpoint should extend far enough into the future to capture the delayed dynamics of interest. At this stage, the reference modes of behaviour are drawn and refer to graphical or verbal descriptions of the historical behaviour of key concepts and variables, including expectations about their future development (Ford, 2010; Randers, 1980; Saeed, 1992). The main objective of this step is to correctly define the problem, ensuring clear and consistent model boundaries that maintain the fit for purpose while seeking for simplicity.

The second step concerns the formulation of a dynamic hypothesis aimed at explaining the underlying causes of the problematic behaviour. This involves identifying the feedback loops and stock–flow relationships that compose the system, as described in Sect. 3.2.3. In this phase, the modeller should seek endogenous explanations by adopting model boundaries broad enough to include all critical causal linkages and interactions among variables. The final output of this step is a draft map of the system structure, which defines what the model includes and excludes. This initial version may be expressed through qualitative System Dynamics tools, such as CLDs or qualitative SFDs (Homer, 2019).

Once the hypothetical system structure is defined, the third step is model formulation, developed in Sect. 3.3 and detailed in the "Appendix B". Here, the literature reviewed in Chap. 2 is reinterpreted through a System Dynamics lens to highlight causal linkages between variables and provide numerical estimations. At this stage, the dynamic hypothesis is tested, moving "from the conceptual realm of diagrams to a fully specified formal model, complete with equations, parameters, and initial conditions" (Sterman, 2000, p. 103). Model formulation also marks the beginning of the testing process, which unfolds continuously through formal and informal practices to ensure consistency throughout the development phase.

The fourth step is model testing, aimed at building confidence and trust among potential users. This involves a variety of tests—including tests of model structure, behaviour, and learning (Morecroft, 2015, pp. 408–410; Sterman, 2000, pp. 859–861)—to assess the quality of both the model and the modelling procedure (Forrester & Senge, 1980). Iterative trials, adjustments, and validations—described in the "Appendix A"—are conducted to refine the model and confirm its robustness.

Once a satisfactory level of confidence has been reached, the researcher can proceed with the fifth and final step: policy formulation and evaluation, which will be addressed in Chap. 4. This step is closely linked to the testing phase, as the analysis of the model's response to policies and parameter changes serves both as a learning and a behavioural test.

As anticipated, the following Sections 3.2 and 3.3 describe the development of the Sustainability Reporting Model. These sections provide essential insights into the model's structure and the rationale behind the behaviours that will be analysed in Chap. 4.

3.2 Preliminary Modelling Steps: Problem Articulation and Dynamic Hypothesis

3.2.1 Problem Articulation: Greenwashing as Emerging Phenomenon

As discussed in Chap. 1, the evolution of the sustainability discourse has not followed a linear path but has been the result of an ongoing dialogue between global actors, characterised by the reaching of agreements that have guided policies and collective choices. In this context, the role of the companies has grown in importance, with the sustainability accounting discipline that has contextually evolved, offering tools and guidelines to disclose performance across the sustainability dimensions.

However, despite the academic and standard setters' efforts to promote sustainability reporting as a comprehensive tool full of managerial implications, organisations still struggle to perceive it as a source of change. This is because the dominant economic paradigm stimulates the protection of business-as-usual practices, the sustainable transition is costly, and the benefits are still uncertain (Calabrese et al., 2022; Sachs, 2012). For this reason, sustainability reporting is often relegated to a mere narrative primarily used to preserve legitimacy rather than to guide transformation. This tendency is exacerbated by the fragmentation of reporting standards and the coexistence of divergent materiality definitions, which lead to heterogeneous applications and hinder comparability (see Sections 2.1 and 2.3).

In this context, materiality is often not integrated into decision-making processes, remaining confined to the reporting domain (see Sect. 2.3.2). Although the consequences may not be immediately visible from the outside, the misalignment between external stakeholder expectations and the internal role attributed to sustainability disclosures can generate several reporting distortions. One of the most common and relevant is the use of sustainability communication to highlight only the positive aspects of business activities, omitting critical issues or negative externalities. This selective representation gives rise to practices that can be defined as greenwashing (Jones, 2019).

Greenwashing, therefore, emerges as a systemic phenomenon rooted in the strategic use of sustainability narratives, not supported by a contextual revision of business models and internal activities. The widespread use of greenwashing techniques undermines the credibility of reporting, leading to a generalised sense of scepticism among stakeholders and consumers. This scepticism, in turn, affects the perceived legitimacy of corporate commitments and may compromise the effectiveness of sustainability strategies over time.

To understand the relevance and evolution of this phenomenon, a preliminary empirical observation was conducted using Google Trends. The search frequency of the term greenwashing was analysed over the period 01/01/2015–30/09/2025. As shown in Fig. 3.1, the data reveal a clear upward trend, with a notable acceleration around 2022. This increase can be interpreted as a proxy for the rise in public awareness: When users search for greenwashing, they are implicitly acknowledging that some of the information they receive may be misleading. They seek to understand the meaning of the term and the implications of unfaithful communication practices. In this sense, search volume becomes an indirect indicator of stakeholder scepticism.

In summary, greenwashing emerges as a strategic response to the growing pressure for sustainability, often disconnected from substantive organisational change. Its diffusion reflects a broader difficulty in integrating sustainability into core business processes and reveals the fragility of current reporting practices when used primarily for reputational purposes.

The increasing public awareness of misleading communication, captured through the rising search interest in the term greenwashing, signals a shift in stakeholder expectations and a growing demand for transparency and accountability.

This evolving dynamic, rooted in the misalignments between external reporting purposes and internal managerial inertia, constitutes the core problem addressed by the model. The next section introduces the reference mode of behaviour and defines the boundaries of the simulation, aiming to capture the long-term consequences of greenwashing and stakeholder scepticism on corporate sustainability performance.

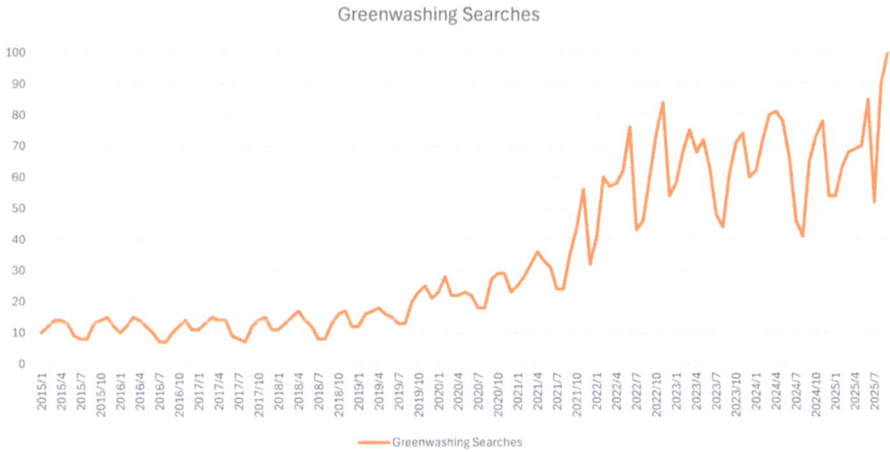

Fig. 3.1 Number of worldwide Google searches for the term "greenwashing" for the period 01/01/2015–30/09/2025. Legenda: The values on the Y-axis represent the percentage of worldwide searches for the term "greenwashing" relative to the peak search volume recorded in September 2025 (indexed as 100). Source: Author's elaboration of Google Trends data. Data extracted on 13/10/2025 and publicly available at: https://trends.google.it/trends/explore?date=2015-01-01%20 2025-09-30&q=greenwashing&hl=it

3.2.2 Reference Mode of Behaviour and Boundary Selection

Tracing a system reference mode of behaviour represents the first step of the modelling process. It consists in describing how the problem evolves over time, based on empirical observations and plausible expectations (Ford, 2010; Randers, 1980; Saeed, 1992). In this research, the reference mode is constructed around the perceived diffusion of greenwashing and the resulting stakeholder scepticism, starting from data shown in Fig. 3.1 and projecting these into the future.

Despite the growing awareness of misleading business practices, identifying measures of scepticism in absolute or relative terms remains challenging. Academic research has rarely focused on quantifying perceived greenwashing, but institutional sources offer useful insights. A screening conducted by the European Commission and national consumer authorities in early 2021 found that 42% of sustainability-related claims could reasonably be considered exaggerated, false, or deceptive, potentially qualifying as unfair commercial practices under EU rules. Although this data does not represent a precise quantification of greenwashing in sustainability reporting, it confirms the widespread nature of the phenomenon.

Similarly, the study by Torelli et al. (2020b) explored stakeholder perceptions of corporate communication practices. Their findings suggest a significant level of perceived greenwashing, with average scores of 4.28 and 4.62 out of 7 on questions related to hidden intentions and exaggerated environmental claims. These values reflect qualitative perceptions rather than measurable proportions of misleading content and therefore can only offer a general idea of the importance of the greenwashing phenomenon.

Based on these observations and the trend shown in Figs. 3.1, and 3.2 presents a stylised evolution of the perceived greenwashing variable, also attributing an ideal value to perceived greenwashing in the form of a percentage of the total sustainability information received by a user. The initial sharp increase is expected to give way to a stabilised and widespread sense of scepticism, which may persist for

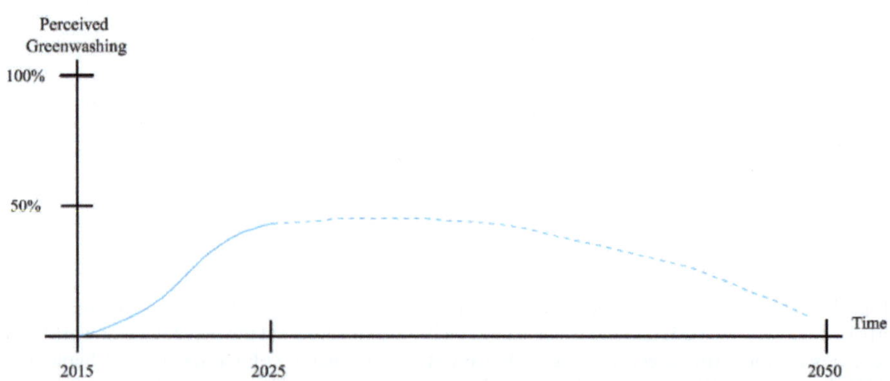

Fig. 3.2 Perceived greenwashing variable: Reference mode of behaviour. Source: Author's elaboration

several years (Torelliet al., 2020b). A gradual decline is hypothesised in the final part of the simulation period, driven by the growing relevance of sustainability issues and the implementation of regulatory measures aimed at discouraging greenwashing practices (European Commission, 2020).

As concerns the time horizon selected, Fig. 3.2 shows how the simulation spans from 2015 to 2050, aligning it to two symbolic milestones: the launch of the SDGs in 2015, which marked global recognition of sustainability challenges, and the EU's Net Zero target in 2050, which represents the expected culmination of the sustainability transition. This period is long enough to explore both the emergence of greenwashing and the long-term consequences of stakeholder scepticism.

For what concerns the model's boundaries, the generic structure of the model represents a simplified analysis of the relationship between a company and its stakeholders (represented as a unique actor). The company can convey information to the outside through its sustainability report, which can contain true information on performance, as well as greenwashing elements. Stakeholders compare the levels of performance disclosed to their expectations and apply filters based on the credibility of the company (i.e., scepticism). Their level of satisfaction has consequences on both the impact (represented by stakeholders' materiality) and financial (represented by financial achievements) dimensions of sustainability. Then, the company internalises this knowledge through materiality assessment.

As discussed in Sect. 2.2, this book adopts an internal perspective of materiality. This means that the outcome of the materiality assessment is treated as a driver of internal change and investment decisions, rather than a mere determinant of report contents. Report contents are assumed to be dependent on external levels of expectations and pressure, in a hypothetical context of regulatory obligation. Substantially, if stakeholder expectations are greater than zero while internal materiality is zero, the resulting information gap generates stakeholder dissatisfaction. In the model, this is represented by treating non-disclosure as the worst level of communicated performance. This reflects the increasing pressure for transparency and the evolving role of reporting as an accountability tool (Abhayawansa, 2022).

All the variables contained in the model are determined endogenously, except for one. This is the average global materiality, which represents the growing relevance of sustainability issues in public discourse and institutional agendas. In the model, this variable acts as an exogenous and continuous driver of change, influencing both stakeholder expectations and the financial relevance of sustainability topics.

To maintain analytical simplicity, the model uses positive values only, meaning that negative performance (e.g., environmental damage) is represented as zero, while misleading communication is modelled as an increased positive disclosure. This allows the simulation of greenwashing effects without introducing negative numerical inputs, while preserving the interpretive clarity of the results.

Soft variables are quantified through simplified mathematical relationships, based on the literature reviewed in Chap. 2, aiming at exploring trends and long-term consequences of alternative strategic choices rather than precise forecasts (Winch et al., 1999).

A summary of the problem articulation and boundary selection is provided in Table 3.1, following Sterman's (2000) guidelines. This table outlines the thematic focus, key variables, time horizon, and modelling assumptions, offering a concise overview of the conceptual framework that supports the simulation.

Table 3.1 The problem articulation and boundary selection step of the modelling process

Theme selection	Sustainability performance measurement and reporting are fundamental to supporting companies in their transition towards sustainable business models. Nevertheless, greenwashing practices are undermining the credibility of the disclosures and may delay this transition. The managerial impacts of alternative approaches to materiality assessment need to be analysed, as well as the existence of leverage points to improve long-term performance.
Key variables	– Performance measures reflect the progress of the sustainability transition. – Perceived greenwashing determines the stakeholders' distrust in business communication. – Stakeholders' satisfaction describes the success of the sustainability strategy. – Financial achievements represent the financial repercussions linked to sustainability performance. – Materiality assessment helps internalise the strategic and financial relevance of sustainability issues.
Time horizon	Start: 2015—Launch of the Agenda 2030, marking global commitment to sustainability. End: 2050—Target year for net zero emissions under the EU green Deal, representing the expected culmination of the sustainability transition.
Dynamic problem definition	Greenwashing is the primary challenge to the credibility and comparability of sustainability measures and communication. Its fast growth in the last years is undermining the effectiveness of sustainability reporting, causing the rise of scepticism. In the future, greenwashing is not expected to grow at the same speed, but scepticism may be maintained for several years.
Model boundaries (setting)	– Setting: The relationship between a company and its stakeholders. The general importance of sustainability issues, along with the potential market dimension, is assumed to grow exogenously. – Case independence: The model represents the general configuration of a company's strategic and reporting approaches to sustainability. – Financial repercussion: The model attributes a generic score (ranging between a minimum of 0 and a maximum of 4) to the financial consequences of sustainability. – Soft modelling: Mathematical relationships are estimated for intangible aspects. The model does not aim to provide precise numerical forecasts of future states of the system but to evaluate trends and long-term repercussions of alternative scenarios and managerial decisions.
Model boundaries (variables)	– Main endogenous variables: Internal materiality, sustainability-dedicated assets and competencies, stakeholders' expected performance, achieved performance; perceived greenwashing, stakeholders' satisfaction, stakeholders' materiality, financial achievements – Exogenous variable: Average global materiality – Variables excluded from the model scope: Behaviour of the company's competitors, sustainability market competition, sense of scepticism due to external conditions, effects of assurance requirements, negative consequences of sustainability performance

Source: Author's elaboration

3.2.3 *Causal Loop Diagram Rooted in the Literature and Theories*

3.2.3.1 A Systemic View of the Sustainability Reporting Theories

The review of the literature conducted in Chap. 2 framed sustainability reporting as a complex and interconnected system. The analysis began with the materiality assessment (Sect. 2.1), disclosing how its impact, financial, and strategic components shape a company's internal level of priority attributed to a sustainability issue. In this way, internal materiality determines how resources are allocated and how performance is managed (Sect. 2.2). The disclosure of sustainability performance takes place in the sustainability report. The role attributed to this document has important implications in the type of information disclosed and the implications in terms of stakeholders' engagement and financial repercussions. Various theories (analysed in Table 2.1, Section 2.3.1) have been used to explain the different approaches, in terms of how external pressures are managed and/or internalised, subsequently shaping stakeholders' responses and defining the impact dimension of materiality (Section 2.4). Furthermore, these interactions determine the financial repercussions of sustainability efforts, which are captured by the concept of financial materiality (Section 2.5).

A comprehensive view of this cycle is offered in Fig. 3.3, which summarises the interconnected relationships into a unified model, locating the different theories along a continuous informative flow that represents the entire sustainability reporting system.

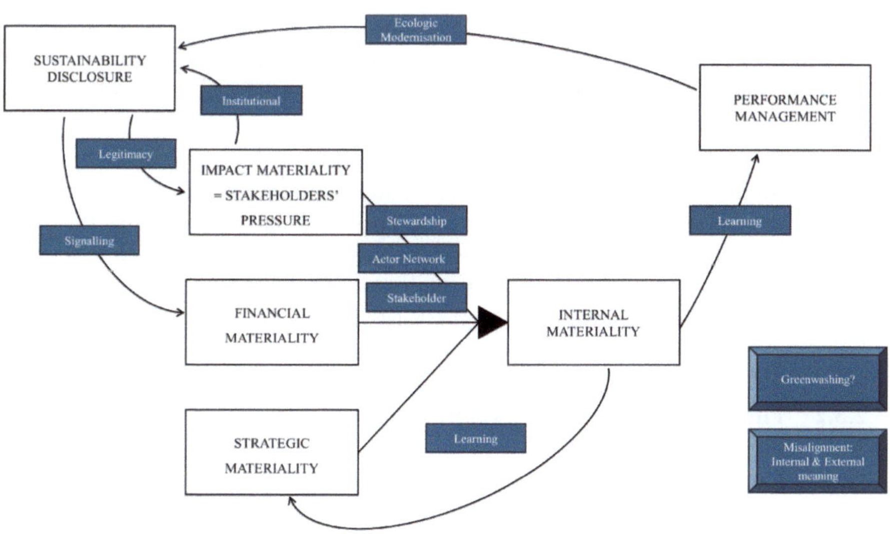

Fig. 3.3 Sustainability reporting as a system. Source: Author's elaboration

Starting from the materiality assessment, *stewardship* and *actor-network* theories describe how organisations internalise social and environmental instances (i.e., impact materiality) into strategy; *stakeholder* theory expands the information flow to include also financial aspects since investors constitute a category of stakeholders. The mutual mediation between internal materiality (i.e., the output of the new assessment) and the previous beliefs (i.e., strategic materiality) are addressed by *learning* theory, in a continuous process of self-improvement. Completing the *learning* process, the level of internal materiality influences resource allocation, enhancing performance management practices. The link resources-performance-disclosure is supported by *ecologic modernisation* theory, for which sustainability transition should be achieved in the current economic system by seeking efficiency and promoting investments.

Several roles can be attributed to sustainability disclosure, which can be summarised in two macro-categories: obtaining the external *legitimacy* of stakeholders or *signalling* the company's commitment to sustainability. Although these two theories can be used to explain both relationships with stakeholders and with the market, Fig. 3.3 attributes a higher impact-oriented meaning to legitimacy aspects, while signalling aspects are mainly linked to competitive advantage and financial repercussions. Finally, the *institutional* theory justifies the consequences of stakeholders' and external pressure on reporting practices.

The acquired holistic view underlines the dynamic interplay between internal processes, external perceptions, and materiality dimensions, contributing to achieving the overall picture of the sustainability reporting phenomenon.

As already seen in Sect. 2.3.3, this sustainability reporting cycle has the potential to promote the sustainability transition of companies, but can also generate undesirable outcomes, represented by untruthful communication—identified in this book as *greenwashing practices*. These constitute the dynamic problem object of investigation (see Sections 3.2.1 and 3.2.2) with scepticism rise and the loss of any managerial purpose of the reporting process that generate dynamic complexity.

For this reason, the two blocks positioned at the margin of Fig. 3.3 ask how can *greenwashing practices* and *misalignments between the internal and external meaning of reporting* be integrated into the modelling process?

The System Dynamics approach offers the means to investigate the cause-effect relationships among the components of the reporting system. In particular, the problem orientation of the discipline mandates the identification of the systemic roots of the problem, which can be identified through the support of CLDs, which includes the polarity of the relationships among the variables.

3.2.3.2 The Sustainability Reporting Causal Loop Diagram

To operationalise this problem-oriented perspective, this section introduces a simplified CLD that offers a synthesis of the sustainability reporting model systemic structure. By mapping the polarity and feedback mechanisms among key variables, the CLD provides a visual representation of the dynamics at play, offering a first interpretive lens to explore the behavioural patterns that will be analysed in the simulation. Figure 3.4 illustrates this structure.

The first dynamic discussed is determined by an exogenously driven change in *stakeholders' materiality*, representing, for example, increasing concerns about a sustainability issue such as climate change or workers' conditions. These concerns raise the benchmark of *stakeholders' expectations* which, in the absence of a company response, decreases the level of *stakeholders' satisfaction*, thereby increasing stakeholders' concerns and, consequently, *stakeholders' materiality*. This self-reinforcing process describes the dynamic for which unaddressed stakeholder pressures grow in magnitude over time, leading to dissatisfaction (Lazarus & Folkman, 1984). The reinforcing feedback loop *R1: stakeholders' pressure process*, depicts this dynamic.

As can be seen in Fig. 3.4, *stakeholders' materiality* also represents the *impact* dimension of *materiality*. For this reason, the company should address the increased magnitude of the analysed sustainability issue through the materiality assessment. In this way, external pressure should lead to an increase in the company's *internal materiality*, which in turn stimulates *investments* aimed at improving sustainability *achieved performance*. This improvement leads to disclosures more aligned with stakeholders' expectations, reducing the gap between expected and disclosed performance. In the CLD, this process is represented by the level of *disclosed performance* positively impacting *stakeholders' satisfaction*, which should then reduce *stakeholders' materiality*, that is, the original source of pressure. This balancing process is described by the balancing loop *B1: company's answer to pressure*.

Over time, increased levels of *stakeholders' satisfaction* should promote positive *financial performance*, which also represents the *financial* dimension of *materiality*.

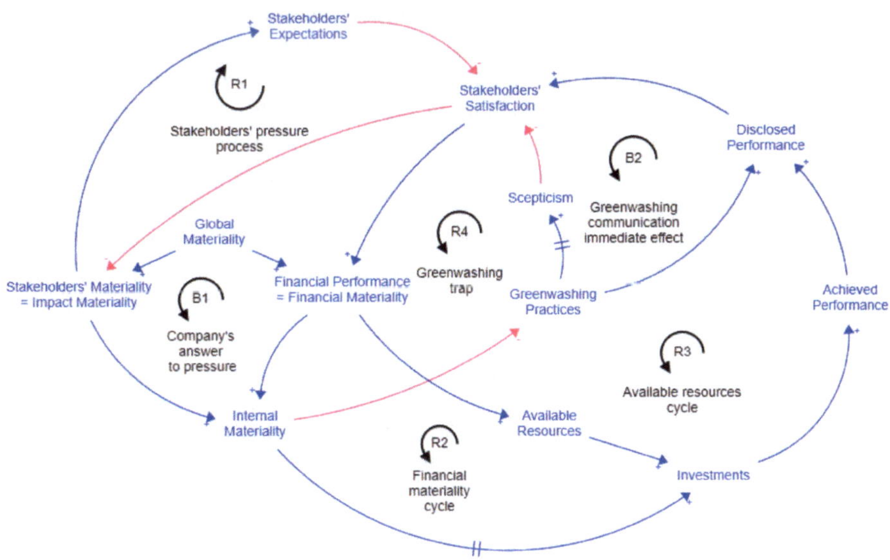

Fig. 3.4 CLD analysis of the model's feedback loops. Source: Author's elaboration

Higher *financial performance* stimulates two virtuous cycles: the first—*R2: financial materiality cycle*—is linked to an increase in the financial component of *internal materiality*, encouraging the company's prioritisation of the sustainability topic; the second—*R3: available resources cycle*—is linked to the new financial *available resources* that can be reinvested to further improve sustainability performance.

As the reader may notice, many conditional verbs are used in the description of the first feedback loops. Indeed, the dynamics described by *R1*, *B1*, *R2*, and *R3* represent an ideal cycle of a company that decides to improve its sustainability performance and invest in the transition. Nevertheless, greenwashing practices exist and must be taken into consideration, as they have spread to the point of becoming a systemic problem capable of interfering with normal dynamics.

For this reason, *greenwashing practices* are placed at the heart of the CLD, creating an unmanageable number of paths that the causal flow can follow and, consequently, new feedback loops with opposite effects. For simplification, only the two major loops (i.e., *B2* and *R4*) are represented and described after the explanation of greenwashing causes and consequences.

The literature review in Sect. 2.3.3 showed how greenwashing is rooted in low levels of business commitment towards sustainability issues; therefore, *greenwashing practices* are modelled as negatively dependent on the level of *internal materiality*[1] (Maniora, 2018; Torelli et al., 2020a). These practices artificially increase disclosed performance, leading to a short-term rise in *stakeholders' satisfaction*, but also to a delayed increase in *scepticism* and loss of trust, which negatively impacts the same *stakeholders' satisfaction* variable (Calabrese et al., 2015; La Torre et al., 2020; Torelli et al., 2020b).

Delayed effects of opposite signs generate complexity in the mental analysis of the system (Barlas, 2009). On the one hand, *greenwashing practices* should self-limit as the company begins to perceive the importance of the sustainability issue, as expressed by the *B2: greenwashing communication immediate effect* loop. This loop represents the dynamics generated by an initial increase in *greenwashing practices* that, by fostering *stakeholders' satisfaction*, eventually reduce the same practices due to an increase in *internal materiality* driven by positive *financial performance*. On the other hand, increases in *greenwashing practices* also cause rises in *scepticism*, which permanently damage *stakeholders' satisfaction*, discouraging *investments* in assets, since the communication would no longer be trusted by stakeholders. In this way, persistent low levels of *stakeholders' satisfaction* limit *financial returns*, preventing the increase of *internal materiality*, and thus

[1] Although greenwashing practices are modelled as negatively dependent on internal materiality, it is important to clarify that the model assumes a context in which sustainability performance is disclosed regardless of the company's internal prioritisation. Further information on this modelling assumption is available in Sect. 3.3.2. Indeed, this book adopts an internal perspective of materiality, treating it as a driver of strategic and investment decisions rather than a determinant of report contents. In this framework, even low levels of internal materiality do not prevent disclosure, which is instead triggered by external stakeholder expectations. Consequently, misleading communication may occur even when the company does not genuinely prioritise the sustainability issue, reinforcing the greenwashing dynamics described in loops *B2* and *R4*.

Table 3.2 Feedback loops in Fig. 3.4

Loop title	Process description
R1: Stakeholders' pressure process	Unaddressed stakeholders' pressures increase in magnitude over time, leading to dissatisfaction.
R2: Financial materiality cycle	Improvements in the achieved sustainability performance increase the financial materiality of the issue, stimulating further investments.
R3: Available resources cycle	Increased financial returns expand the pool of available resources that can be reinvested in sustainability-dedicated assets and competencies.
R4: Greenwashing trap	Greenwashing communication raises scepticism, decreasing satisfaction and financial performance. This discourages investments and promotes further greenwashing.
B1: Company's answer to pressure	Improvements in sustainability achieved performance reduce stakeholder concerns, balancing the source of external pressure.
B2: Greenwashing communication immediate effect	Greenwashing boosts short-term satisfaction and financial performance. The increase in financial materiality may guide the company towards genuine sustainability investments, reducing greenwashing practices.

Source: Author's elaboration

promoting greenwashing despite investment, giving rise to the reinforcing loop *R4: greenwashing trap*, which is responsible for this dynamic.

The feedback loops depicted in Fig. 3.4 and described above are also synthesised in Table 3.2. Each loop is labelled according to its polarity and thematic focus, offering a concise overview of the systemic behaviours that shape the sustainability reporting process.

Albeit only the most representative loops have been described, the structure of the system suggests that different dynamics may emerge depending on which loop becomes dominant over time (Richardson, 1995). Using the train tracks as a metaphor, the level of *internal materiality* can act as a turnout (i.e., a switch) capable of diverting the business strategy from a greenwashing approach towards a genuine commitment to sustainability. The benefits of qualitative modelling (i.e., CLDs) lie in the identification of these feedback structures (Homer & Oliva, 2001). However, mathematical modelling is essential to simulate the theorised dynamics and analyse the most influential points of leverage to govern the system and orient behaviour towards desirable trajectories (Kunc et al., 2021; Richardson, 1997). For this reason, the next section focuses on the description of the quantitative model formulation.

3.3 Model Formulation: Translating Literature into a Generic Quantitative Model

After having identified the hypothetical system structure, this section focuses on the conversion of the logical relationships among variables into mathematical terms, through the definition of equations, parameters, and initial conditions of stocks

(Sterman, 2000, p. 103). The structure presented here is still a simplified representation of the model, articulated through Stock and Flow Diagrams. The complete mathematical formulation, including auxiliary variables and simulation equations, is detailed in Appendix B″.

The aim of this section is to formalise the Sustainability Reporting Model by describing the five interconnected sectors that compose its architecture. Each sector represents a key dimension of the reporting cycle and is described in turn, highlighting the causal relationships and feedback mechanisms that shape system behaviour:

- Internal materiality sector—Sect. 3.3.1
- Resource allocation sector—Sect. 3.3.2
- Performance communication sector—Sect. 3.3.3
- Impact materiality sector—Sect. 3.3.4
- Financial materiality sector—Sect. 3.3.5
- The complete model—Sect. 3.3.6

The division of the model into sectors supports analytical clarity and simulation robustness. Each sector has been independently tested, refined, and validated before integration into the full model.

3.3.1 Sector 1: The Internal Materiality Sector

The first sector of the model, named "internal materiality sector", offers a System Dynamics representation of the internal materiality assessment, that is, the moment in which a company determines the level of priority sustainability issue (OIBR, 2022). From a System Dynamics perspective, the literature elements of Sect. 2.1 can be summarised as follows:

1. Internal materiality
   (a) Type: Stock soft variable
   (b) Range: [0, 1] materiality units
   (c) Description: Internal materiality reflects the accumulated priority attributed to a sustainability issue over time (Ferrero-Ferrero et al., 2018). Its evolution is governed by the flow "Change in internal materiality". This flow is influenced both by the outcomes of the materiality assessment, which includes impact and financial dimensions, and by the company's system of values and beliefs, represented by the previous level of internal materiality (OIBR, 2022).
   (d) Initial condition: The simulation assumes a low initial value (0.10), indicating marginal relevance of the issue.

2. Cumulative effect on internal materiality
   (a) Type: Auxiliary variable
   (b) Description: It reproduces the influence of both impact and financial materiality on the company's internal prioritisation. It is calculated through a weighted sum of two components: the effect of financial materiality and the effect of impact materiality (European Commission, 2023; Stoughton & Ludema, 2012).
3. Time to adjust internal materiality
   (a) Type: Delay parameter
   (b) Description: This parameter represents the time required to revise the internal materiality assessment. It accounts for the organisational inertia and the periodic nature of strategic reviews. Although reporting standards often require annual assessments, empirical evidence suggests that companies perform them on average every 1.4 years, with longer intervals when strategic revisions are involved (Garst et al., 2022; Morecroft, 2015). For this, it is set to 2 years.
4. Weight of financial materiality
   (a) Type: Auxiliary parameter
   (b) Description: This variable determines the relative importance attributed to financial materiality in the cumulative effect calculation. A value of 0.5 simulates a double materiality approach, while values closer to 0 or 1 reflect impact- or financial-oriented perspectives, respectively. It allows the model to represent different strategic orientations and reporting philosophies (Galpin et al., 2015; van der Waal & Thijssens, 2020).

Figure 3.5 provides a visual representation of the causal structure described above.

3.3.2 Sector 2: The Resource Allocation Sector

The second sector of the model, named "resource allocation sector", describes how internal materiality influences the allocation of financial resources and the development of sustainability-related assets and competencies. As discussed in Sect. 2.2, internal materiality represents the starting point for aligning business strategy with stakeholders' and shareholders' priorities (Cerbone & Maroun, 2020; Font et al., 2016). An increase in internal materiality promotes the allocation of resources to acquire tangible assets and develop specific competencies, ultimately improving sustainability performance and its disclosure (Calabrese et al., 2016; Bianchi, 2016).

This is expressed in the model as follows:

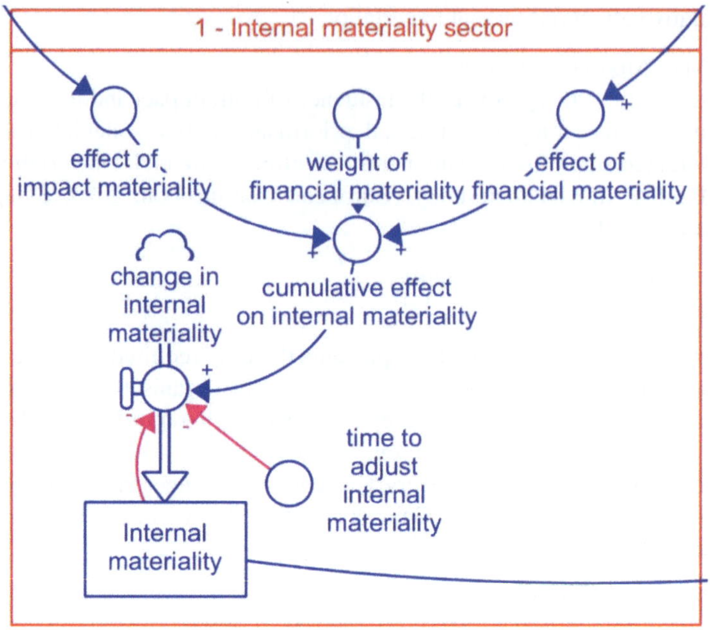

Fig. 3.5 Stock and flow diagram n. 1: The internal materiality sector. Source: Author's elaboration

1. Relative internal materiality
   (a) Type: Auxiliary variable
   (b) Description: It represents the ratio between the current level of internal materiality and a predefined threshold. It determines whether the sustainability issue is considered strategically relevant enough to justify resource allocation (Cerbone & Maroun, 2020; Font et al., 2016).
2. Internal materiality threshold
   (a) Type: Parameter
   (b) Description: It defines the minimum level of internal materiality required for the company to consider a sustainability issue worthy of investment (Calabrese et al., 2016). A neutral value of 0.5 is used to simulate standard strategic behaviour.

Once the level of internal materiality is compared to a predefined threshold, the company faces two strategic alternatives to answer external pressures: It can allocate resources to improve sustainability performance or, alternatively, recur to greenwashing practices (by, for example, exaggerating communications or hiding negative externalities). These two strategic options are presented in turn, starting from the investment pathway:

3. Sustainability-dedicated assets and competencies.
   (a) Type: Stock variable
   (b) Range: > 0 monetary units
   (c) Description: This stock represents the tangible and intangible resources that a company dedicates to sustainability. It includes physical assets (e.g., green technologies, infrastructure) and immaterial competencies (e.g., specialised knowledge, organisational capabilities) that contribute to the achievement of sustainability performance (Bianchi, 2016; Kaplan & Norton, 1992; Yusliza et al., 2020). The stock is subject to depletion over time, reflecting asset obsolescence and knowledge decay (Johnson et al., 2017; Kianto et al., 2014). As for the initial condition of the stock, the simulation assumes a low initial value (1 monetary unit), representing a minimal initial endowment dedicated to sustainability.

The stock increases through two distinct inflows—"available resource allocation" and "reinvested financial achievements"—depending on:

4. Available resources
   (a) Type: Stock variable
   (b) Range: > 0 monetary units
   (c) Description: It represents the reserves already available to the company for potential investment in sustainability. These resources are allocated based on the effect of internal materiality, which can speed up or slow down a predetermined investment time. In initial settings, an amount of available resources of 10 monetary units is allocated according to a 0.25 normal percentage of resources factor in an investment time of 3 years, recreating a situation in which, supposed the level of materiality equal to the threshold (i.e., effect of internal materiality = 1), a company may invest the 25% of its available resources during a 3-year business plan.
5. Financial achievements: A variable belonging to Sector 5 and representing the financial repercussions of sustainability performance

   In both cases, resource allocation is influenced by the variable:

6. Effect of internal materiality on resource allocation
   (a) Type: Auxiliary variable
   (b) Description: It determines how internal materiality influences resource allocation. The effect is neutral when internal materiality equals the threshold and increases or decreases accordingly.

The alternative strategic pathway is based on greenwashing—that is, an approach aimed at satisfying stakeholder expectations without implementing real changes. This behaviour is often rooted in low levels of internal materiality, which is captured by the following model dynamics:

7. Effect of internal materiality on greenwashing
   (a) Type: Auxiliary variable
   (b) Description: It determines how internal materiality influences the greenwashing incentive. The relationship is inversely proportional: higher internal materiality reduces the likelihood of greenwashing, while lower levels increase it (Maniora, 2018; Torelli et al., 2020a).
8. Greenwashing propensity
   (a) Type: Parameter
   (b) Description: This parameter shows the company's system of value concerning greenwashing practices. It acts as a multiplier, with a value of 0 that excludes any form of greenwashing, independently of the level of internal materiality. Initially, its value is set to 0.5, indicating that the 50% of performance shortfall is compensated when the materiality equals the threshold.
9. Greenwashing incentive
   (a) Type: Auxiliary variable
   (b) Range: 0 to 1 (percentage of performance shortfall compensated through greenwashing)
   (c) Description: This variable is the result of the multiplication between the effect of internal materiality on greenwashing and greenwashing propensity. It represents the percentage of performance shortfall compensated through misleading communication (Ferrero-Ferrero et al., 2021; van der Waal & Thijssens, 2020).

The resource allocation sector translates the company's internal prioritisation of sustainability issues into strategic action. By modelling both investment and greenwashing pathways, it captures the dual nature of corporate responses to stakeholder pressure. The level of internal materiality acts as a critical lever, guiding the allocation of financial resources and shaping the credibility of sustainability communication. Figure 3.6 provides a visual representation of the sector's causal structure.

3.3.3 Sector 3: The Performance Communication Sector

The third sector of the model, named "performance communication sector", describes how the company's sustainability performance is communicated and perceived by stakeholders. It models the interaction between achieved and communicated performance, stakeholder expectations, and the effects of greenwashing on stakeholders' satisfaction. This sector is positioned at the intersection between the company's internal efforts and external perceptions and includes both short-term communication dynamics and long-term reputational consequences.

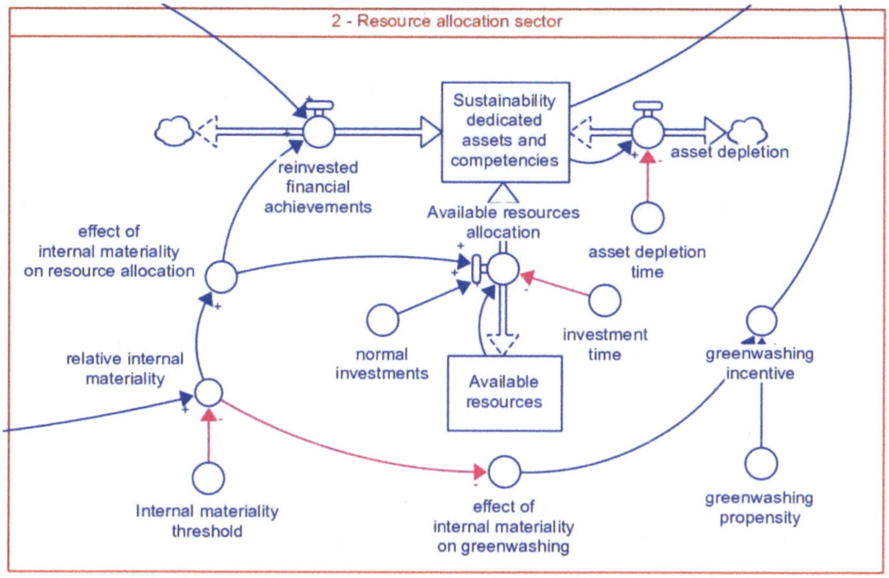

Fig. 3.6 Stock and flow diagram n. 2: The resource allocation sector. Source: Author's elaboration

1. Stakeholders' expected performance
   (a) Type: Auxiliary variable
   (b) Range: [0; 1] performance units
   (c) Description: It represents a dynamic benchmark against which stakeholders evaluate the company's sustainability disclosures (Hoffman & Woody, 2008). It is derived from stakeholders' materiality and reflects their expectations regarding performance (Carroll, 1991; Millar & Searcy, 2020); at the same time, it represents the external pressure exerted by stakeholders (Fernandez-Feijoo et al., 2014; Locke & Latham, 1990).
2. Achieved performance
   (a) Type: Auxiliary variable
   (b) Range: [0; 1] performance units
   (c) Description: It represents the actual performance achieved by the company as a result of investments in sustainability-dedicated assets and competencies. It is influenced by the relative investment level and reflects the company's substantive efforts (Frankel, 1962; Yusliza et al., 2020).
3. Performance shortfall
   (a) Type: Auxiliary variable
   (b) Description: It measures the gap between stakeholders' expected performance and the company's achieved performance. This gap can be left unanswered or compensated through some forms of greenwashing (Maniora, 2018; Torelli et al., 2020a).

4. Greenwashing performance communication
   (a) Type: Auxiliary variable
   (b) Description: It represents the portion of the performance shortfall that is compensated through misleading communication. It is determined by the greenwashing incentive defined in Sector 2.
5. Communicated performance
   (a) Type: Auxiliary variable
   (b) Range: [0; 1] performance units
   (c) Description: It is the sum of achieved performance and greenwashing communication. It represents the total performance disclosed to stakeholders (Delmas & Burbano, 2011).
6. Communication adequacy
   (a) Type: Auxiliary variable
   (b) Description: It measures the percentage of alignment between communicated performance and stakeholders' expectations.

Communication adequacy plays a central role in shaping stakeholders' satisfaction. When the performance disclosed by the company aligns with stakeholder expectations, it contributes to increasing satisfaction levels. However, this effect is not automatic: It is mediated by the perceived credibility of the communication itself. In particular, if stakeholders suspect that the disclosed performance has been exaggerated or manipulated, the positive impact of communication adequacy may be significantly reduced. For this reason, the model introduces the concept of perceived greenwashing, which captures the long-term consequences of scepticism and reputational erosion.

7. Relative greenwashing
   (a) Type: Auxiliary variable
   (b) Description: It is the ratio between greenwashing communication and total communicated performance. It serves as the input for the perceived greenwashing stock.
8. Perceived greenwashing
   (a) Type: Stock variable (third-order delay, see Morecroft, 1983)
   (b) Range: [0; 1] dimensionless
   (c) Description: It represents the level of stakeholder scepticism resulting from misleading communication. It evolves slowly—both in terms of growth and decline—and reflects the delayed effects of reputational damage (Jones, 2019) by modulating the impact of communication adequacy on stakeholders' satisfaction. Higher levels of perceived greenwashing reduce the effectiveness of communication (Lewicki & Bunker, 1996; Kramer, 1999).

9. Perceived greenwashing adjustment time
   (a) Type: Delay parameter
   (b) Description: A long time is required to become aware of greenwashing practices, but once greenwashing is discovered, the loss of trust will impact business behaviour in the long term. For this reason, it is set to 9 years and can be set to a value close to 0 to simulate greenwashing scandals.
10. Communication effect on satisfaction
   (a) Type: Auxiliary variable
   (b) Description: It combines communication adequacy and perceived greenwashing to determine the overall impact of communication on stakeholders' satisfaction (Calabrese et al., 2015).

The communication effect on satisfaction represents the combined influence of what the company communicates and how that communication is perceived. It integrates both the adequacy of disclosed performance and the level of stakeholder scepticism, offering a dynamic measure of how communication contributes to stakeholders' satisfaction. Taken together, these effects shape the overall perception that stakeholders develop regarding the company's approach to the sustainability issue.

11. Stakeholders' satisfaction
   (a) Type: Stock variable (soft variable)
   (b) Range: [0; 1] satisfaction units
   (c) Description: It is the central variable of the sector. It reflects stakeholders' overall perception of the company's sustainability efforts according to the previously described effect (Collins et al., 2007). At the beginning of the simulation, it is set to a value of 0.5, which signals a medium level of satisfaction.
12. Time to adjust satisfaction
   (a) Type: Delay parameter
   (b) Description: It is the required time to change stakeholders' satisfaction level. It is set to 3 years to include brand commitment (or prejudice) and refers to the time needed to change opinion on a firm's overall sustainability behaviour.

In sum, the performance communication sector captures the complex interplay between actual performance, stakeholder expectations, and reputational dynamics. It highlights how greenwashing practices, even when effective in the short term, can undermine stakeholder trust and reduce satisfaction over time. Figure 3.7 provides a visual representation of the sector's causal structure.

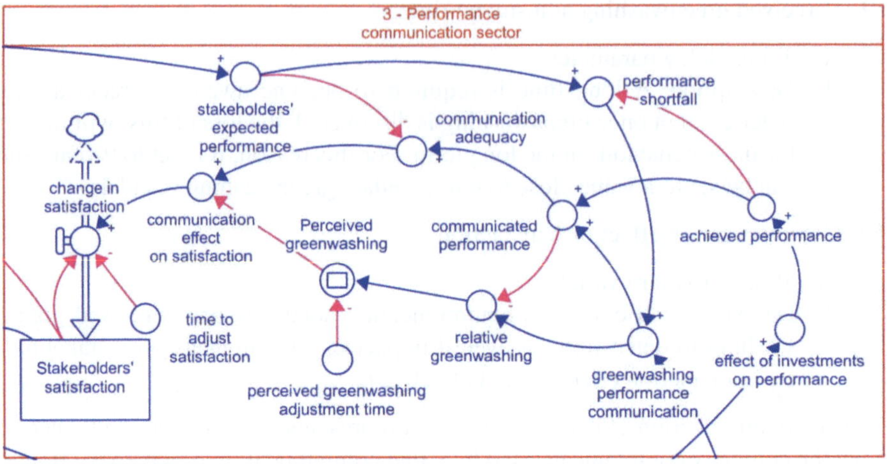

Fig. 3.7 Stock and flow diagram n. 3: The performance communication sector. Source: Author's elaboration

3.3.4 Sector 4: The Impact Materiality Sector

The fourth sector of the model, named "impact materiality sector", focuses on the level of priority that stakeholders attribute to a sustainability issue. This variable, modelled as a stock called stakeholders' materiality, plays a dual role: it acts as the source of external pressure on the company (i.e., stakeholder expectations) (Freiberg et al., 2019; Millar & Searcy, 2020) and as a determinant of impact materiality in the company's internal assessment (OIBR, 2022).

Stakeholders' materiality is influenced by two components: the balancing effect of the company's behaviour (Lazarus & Folkman, 1984), summarised by the level of stakeholders' satisfaction, and the broader societal relevance of the issue, captured by the exogenous variable average global materiality.

1. Stakeholders' materiality
   (a) Type: Stock variable (soft variable)
   (b) Range: [0; 1] materiality units
   (c) Description: It represents the perceived importance of a sustainability issue from the stakeholders' perspective. It evolves over time based on both internal company behaviour and external factors.
2. Time to adjust stakeholders' materiality
   (a) Type: Delay parameter
   (b) Description: It represents the inertia in stakeholder perception. A delay of 2 years is assumed to reflect the time required for stakeholders to revise their views on a sustainability issue.

3. Effect of satisfaction on stakeholders' materiality
   (a) Type: Auxiliary variable
   (b) Description: It captures the balancing effect of stakeholders' satisfaction on materiality. When satisfaction is high, stakeholders reduce their pressure; when low, pressure increases (Lazarus & Folkman, 1984; Calabrese et al., 2019).
4. Average global materiality
   (a) Type: Exogenous variable
   (b) Range: 0 to 1 materiality units
   (c) Description: It represents the societal relevance of the sustainability issue, based on media attention, policy agendas, and global discourse. It increases gradually over time, according to the expected urgency for the sustainability transition (Barth et al., 2017; Di Vaio et al., 2020).
5. Cumulative effect on stakeholders' materiality
   (a) Type: Auxiliary variable
   (b) Description: It combines the effects of stakeholders' satisfaction and global materiality to determine the overall change in stakeholders' materiality. A satisfaction weight parameter allows to change the relative influence of stakeholders' satisfaction versus global materiality in shaping stakeholders' materiality. A value of 0.5 simulates equal weighting (Ferrero-Ferrero et al., 2018).

The impact materiality sector models how stakeholder perceptions evolve in response to both corporate behaviour and societal trends. It provides the foundation for stakeholder expectations and influences the company's internal materiality assessment. Figure 3.8 illustrates the causal structure of this sector.

3.3.5 Sector 5: The Financial Materiality Sector

The fifth sector of the model, named "financial materiality sector", describes how sustainability performance translates into financial outcomes. It simplifies the complex network of causalities into three core variables: the sustainability-related market dimension, the effect of stakeholders' satisfaction on financial achievements, and the resulting financial achievements. These variables are conceptual aggregates, each supported by a set of auxiliary modelling structures that ensure analytical consistency and simulation accuracy (see "Appendix B"). This level of aggregation is necessary to maintain analytical clarity while capturing the essential dynamics of financial materiality, as better discussed in Sect. 2.5.

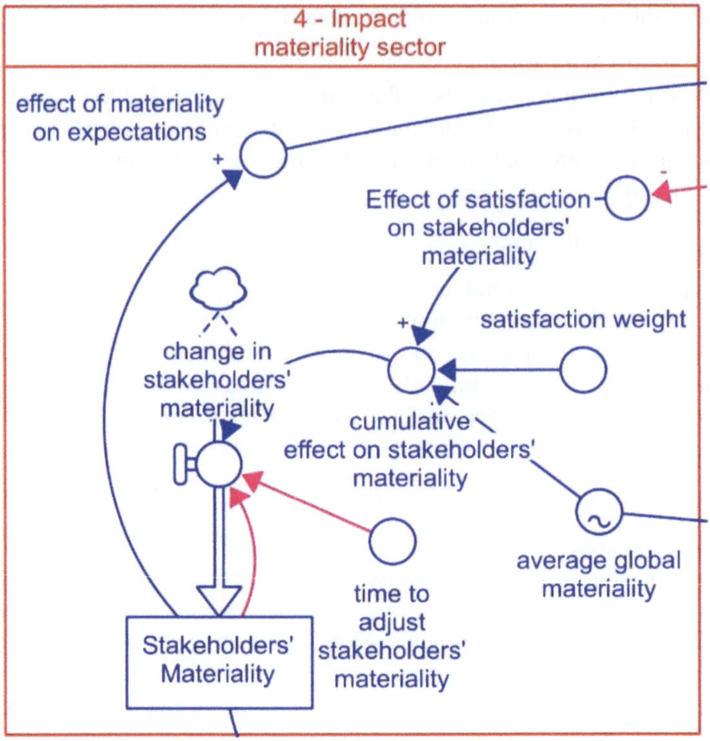

Fig. 3.8 Stock and flow diagram n. 4: The impact materiality sector. Source: Author's elaboration

1. Sustainability-related market dimension
   (a) Type: Auxiliary variable
   (b) Description: Represents the potential financial benefits associated with a sustainability issue. It is influenced by the average global materiality, which reflects the societal and institutional relevance of the issue. Indeed, the higher the global importance of a theme, the higher the amount of money that people may spend on green products or the possibility of taxes or public incentives that directly impact the business dimension (Muñoz-Torres et al., 2019; United Nations, 2023). This determines the potential financial benefits that a company may achieve by completely satisfying stakeholders' expectations (Porter & Kramer, 2006). Its maximum value can be of four units, as better explained in the calibration details in Sect. 3.3.6.

2. Effect of satisfaction on financial achievements
   (a) Type: Auxiliary variable
   (b) Range: [0;1] dimensionless
   (c) Description: Acts as a scaler for the financial benefits achievable. Higher satisfaction levels indicate stronger stakeholder approval, which enhances the company's ability to capture market opportunities and financial incentives (Orlitzky et al., 2003; Waddock & Graves, 1997).
3. Financial achievements/financial materiality
   (a) Type: Auxiliary variable
   (b) Description: Represents the financial returns derived from sustainability performance. It is calculated by multiplying the market potential (based on global materiality) by the effect of satisfaction on financial achievements. This variable also represents the financial materiality of the issue, as it reflects the financial repercussions of the company's sustainability performance (IFRS, 2023; Schiehll & Kolahgar, 2021).

The model assumes that when either of the stakeholders' satisfaction or market relevance is zero, financial materiality also drops to zero, indicating no involvement in a certain sustainability market. Negative financial consequences of poor sustainability performance are excluded from the model boundaries, as previously acknowledged in Table 3.1 (see Sect. 3.2.2).

Financial achievements can be reinvested in sustainability assets, contributing to the resource allocation sector. Alternatively, they may be distributed or redirected, going out of model boundaries. Figure 3.9 illustrates the causal structure of this sector.

The financial materiality sector provides a simplified representation of how sustainability efforts generate financial value. By linking global materiality and

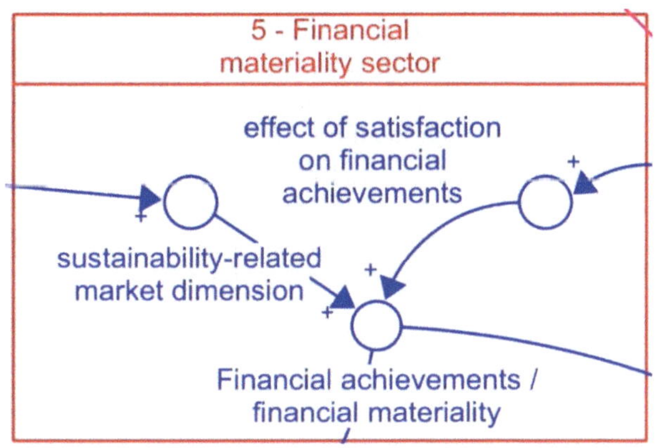

Fig. 3.9 Stock and flow diagram n. 5: The financial materiality sector. Source: Author's elaboration

stakeholders' satisfaction to financial achievements, it closes the feedback loop between performance, perception, and financial outcomes, reinforcing the strategic relevance of sustainability within the business model.

3.3.6 The Sustainability Reporting Model

To conclude the model formulation phase, this subsection presents a comprehensive view of the Sustainability Reporting Model, integrating the five sectors into a unified structure. This complete picture is illustrated in Fig. 3.10.

Additionally, the paragraphs below provide a brief overview of the model's calibration and general mathematical structure. Since the model is generic and not based on a specific case study, the estimation of the resource allocation and performance communication sectors was conducted using an idealised, simplified scenario.

This calibration scenario excludes both greenwashing practices and exogenous increases in global materiality, reducing the determinants of stakeholders' satisfaction and materiality levels to the performance derived from sustainability-dedicated assets and competencies. Internal materiality is assumed to be at a medium level (0.5 out of 1), exactly matching stakeholders' levels of materiality and satisfaction. This setup ensures that internal, impact, and financial materiality are aligned, thereby excluding managerial inputs for changing business behaviour.

To reflect this managerial desire for stability, the reinvestment of financial achievements was calibrated to offset asset depletion, sustaining a constant performance level of 0.25, that is, half of the stakeholders' expectations (0.5), and maintaining stakeholders' satisfaction at a medium level of 0.5.

In this idealised and simplified scenario, equilibrium is maintained throughout the simulation. The parameters that support this configuration are summarised in Table 3.3.

With all essential elements of the model now defined, Chap. 4 will explore its dynamic behaviour through scenario simulations, offering insights for both academic research and managerial decision-making. For readers interested in model validation and reproduction, "Appendix A" contains all relevant information on tests, while "Appendix B" provides equations of the variables and parameter values, enabling replication within System Dynamics software environments.

All modelling procedures, tests, and simulations have been carried out through the software Stella Architect, Version 3.8.

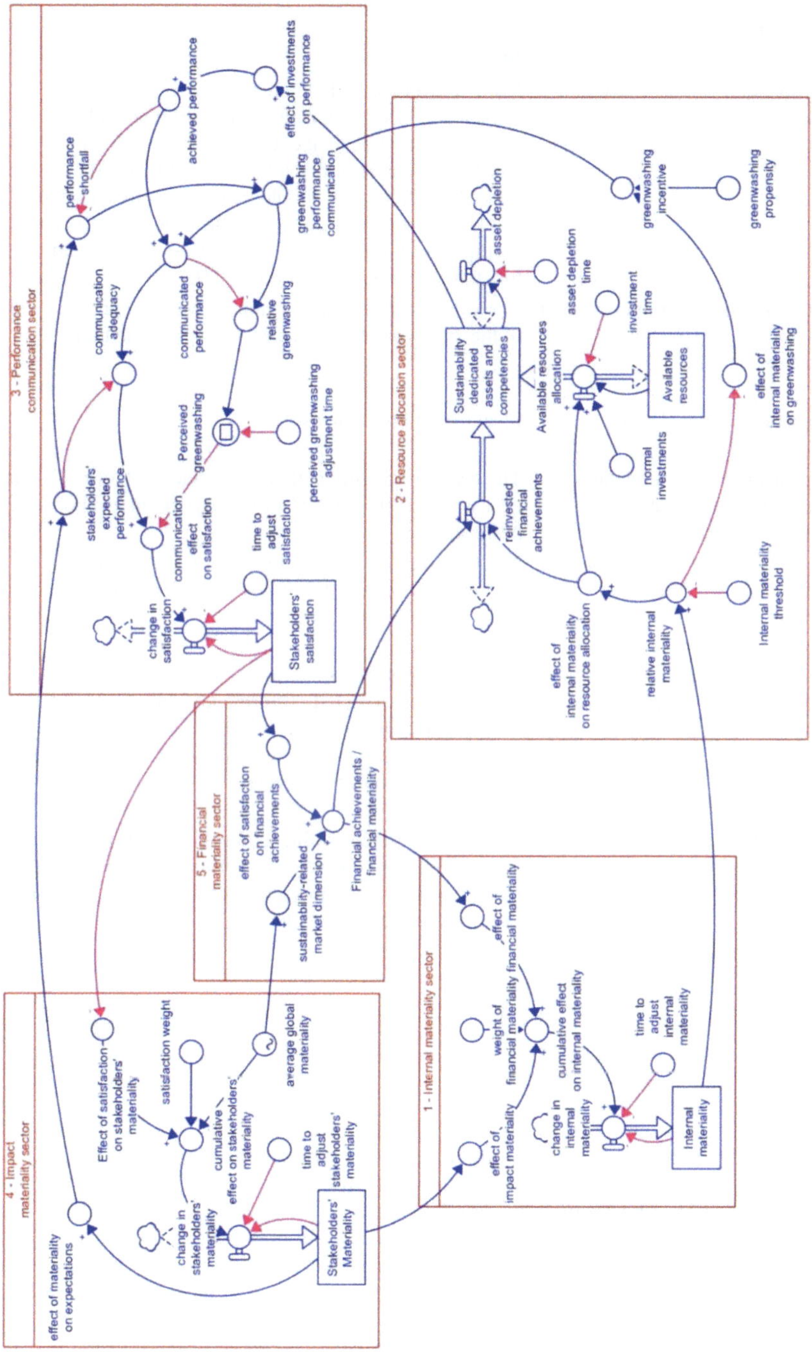

Fig. 3.10 Stock and flow diagram of the entire Sustainability Reporting Model. Source: Author's elaboration

Table 3.3 Parameters estimated through model calibration to an ideal scenario

Parameter name (Sector)	Value, equation, or point	Justification
Effect of financial on internal materiality (sector 1)	(1; 1) point of the table function	Normal levels of global materiality and stakeholders' satisfaction of 0.5 determine a level of financial achievements (i.e., financial materiality) of 1. This means that a level of financial materiality equal to 1 is a normal condition that should be reflected in a normal level of internal materiality. For this, the effect (i.e., the multiplier) should assume a neutral value of 1, with the function passing from the point (1;1).
Initial sustainability assets and competencies (sector 2)	1 monetary unit	The value of the stock of 1 monetary unit represents 4% of the endowment necessary to complete the transition.
Asset depletion time (sector 2)	10 years	Ideal useful life of an asset.
Normal dedicated investment level (sector 3)	5 monetary units	The level of investment that is considered "normal". It is necessary to achieve 25% of the maximum level of performance.
Effect of investments on performance (sector 3)	(1;1) point of the table function	The function passes from the point (1;1), meaning a normal level of performance when investments are at their normal level.
Normal performance level (sector 3)	0.25 performance units	When stakeholders' materiality is at a medium level (i.e., 0.5), the expected performance will also be 0.5. This means that achieving half of it (i.e., 0.25 normal performance) maintains a medium level of satisfaction, keeping stakeholders' expectations at a medium 0.5 level.
Sustainability-- related Market dimension (sector 5)	4	This level of potential financial achievements linked to sustainability performance is calibrated to the resource allocation values. In normal conditions, reinvestments are equal to the asset depletion.

Source: Author's elaboration

References

Abhayawansa, S. (2022). Swimming against the tide: Back to single materiality for sustainability reporting. *Sustainability Accounting, Management and Policy Journal, 13*(6), 1361–1385. https://doi.org/10.1108/SAMPJ-07-2022-0378

Amaduzzi, A. (1953). *L'azienda nel suo sistema e nell'ordine delle sue rilevazioni*. Utet.

Barlas, Y. (2002). System dynamics: Systemic feedback modeling for policy analysis. In *Knowledge for sustainable development–an insight into the encyclopedia of life support systems* (Vol. 1, pp. 1131–1175). UNESCO-Eolss Publishers.

Barlas, Y. (2009). *System dynamics: Systemic feedback modeling for policy analysis* (Vol. 1 and 2). Eolss Publishers Company Limited.

Barth, M. E., Cahan, S. F., Chen, L., & Venter, E. R. (2017). The economic consequences associated with integrated report quality: Capital market and real effects. *Accounting, Organizations and Society, 62*, 43–64. https://doi.org/10.1016/j.aos.2017.08.005

Bianchi, C. (2016). *Dynamic performance management* (Vol. 1). Springer International Publishing. https://doi.org/10.1007/978-3-319-31845-5

Boulding, K. E. (1956). General systems theory–the skeleton of science. *Management Science, 2*(3), 120–131.

Brown, G. S., & Campbell, D. P. (1948). *Principles of servomechanisms dynamics and synthesis of closed loop control systems*. MIT.

Cabrera, D., Colosi, L., & Lobdell, C. (2008). Systems thinking. *Evaluation and Program Planning, 31*(3), 299–310. https://doi.org/10.1016/j.evalprogplan.2007.12.001

Calabrese, A., Costa, R., Levialdi Ghiron, N., & Menichini, T. (2019). Materiality analysis in sustainability reporting: A tool for directing corporate sustainability towards emerging economic, environmental and social opportunities. *Technological and Economic Development of Economy, 25*(5), 1016–1038. https://doi.org/10.3846/tede.2019.10550

Calabrese, A., Costa, R., Levialdi Ghiron, N., Tiburzi, L., & Villazon Montalvan, R. A. (2022). Is the private sector becoming cleaner? Assessing the firm's contribution to the 2030 agenda. *Journal of Cleaner Production, 363*. https://doi.org/10.1016/j.jclepro.2022.132324

Calabrese, A., Costa, R., Levialdi, N., & Menichini, T. (2016). A fuzzy analytic hierarchy process method to support materiality assessment in sustainability reporting. *Journal of Cleaner Production, 121*, 248–264. https://doi.org/10.1016/j.jclepro.2015.12.005

Calabrese, A., Costa, R., & Rosati, F. (2015). A feedback-based model for CSR assessment and materiality analysis. *Accounting Forum, 39*(4), 312–327. https://doi.org/10.1016/j.accfor.2015.06.002

Carroll, A. B. (1991). The pyramid of corporate social responsibility: Toward the moral Management of Organizational Stakeholders. *Business Horizons, 34*, 4–39.

Cerbone, D., & Maroun, W. (2020). Materiality in an integrated reporting setting: Insights using an institutional logics framework. *The British Accounting Review, 52*(3). https://doi.org/10.1016/j.bar.2019.100876

Collins, C. M., Steg, L., & Koning, M. A. S. (2007). Customers' values, beliefs on sustainable corporate performance, and buying behavior. *Psychology and Marketing, 24*(6), 555–577. https://doi.org/10.1002/mar.20173

Coyle, G. (2000). Qualitative and quantitative modelling in system dynamics: Some research questions. *System Dynamics Review, 16*(3), 225–244. https://doi.org/10.1002/1099-1727(200023)16:3<225::AID-SDR195>3.0.CO;2-D

Delmas, M. A., & Burbano, V. C. (2011). The drivers of greenwashing. *California Management Review, 54*(11).

Di Vaio, A., Syriopoulos, T., Alvino, F., & Palladino, R. (2020). "Integrated thinking and reporting" towards sustainable business models: A concise bibliometric analysis. In Meditari Accountancy Research (Vol. 29, Issue 4, pp. 691–719). Emerald Group Holdings Ltd. https://doi.org/10.1108/MEDAR-12-2019-0641.

European Commission. (2020). *Financing the green transition: The European green Deal investment plan and just transition mechanism*. EU Press Release.

European Commission (2023). *Commission Delegated Regulation (EU) 2023/2772 of 31 July 2023 supplementing directive 2013/34/EU of the European Parliament and of the council as regards sustainability reporting standards*.

Fernandez-Feijoo, B., Romero, S., & Ruiz, S. (2014). Effect of stakeholders' pressure on transparency of sustainability reports within the GRI framework. *Journal of Business Ethics, 122*(1), 53–63. https://doi.org/10.1007/s10551-013-1748-5

Ferrero-Ferrero, I., Fernández-Izquierdo, M. Á., Muñoz-Torres, M. J., & Bellés-Colomer, L. (2018). Stakeholder engagement in sustainability reporting in higher education. *International Journal of Sustainability in Higher Education, 19*(2), 313–336. https://doi.org/10.1108/IJSHE-06-2016-0116

Ferrero-Ferrero, I., León, R., & Muñoz-Torres, M. J. (2021). Sustainability materiality matrices in doubt: May prioritizations of aspects overestimate environmental performance? *Journal of Environmental Planning and Management, 64*(3), 432–463. https://doi.org/10.1080/09640568.2020.1766427

Font, X., Guix, M., & Bonilla-Priego, M. J. (2016). Corporate social responsibility in cruising: Using materiality analysis to create shared value. *Tourism Management, 53*, 175–186. https://doi.org/10.1016/j.tourman.2015.10.007

Ford, A. (2010). Modeling the environment (2nd Edition). Island Press.

Forrester, J. W. (1961). *Industrial dynamics*. Massachusetts Institute of Technology (MIT).

Forrester, J. W., & Amaduzzi, A. (1974). *Principi Dei sistemi (principles of systems)*. Etas Kompass.

Forrester, J. W., & Senge, P. M. (1980). Tests for building confidence in system dynamics models. In a. A. Jr. Legasto & R. E. Machol (Eds.), *System dynamics* (Vol. 14, pp. 209–228).

Frankel, M. (1962). The production function in allocation and growth: A synthesis. *The American Economic Review, 52*(5), 996–1022.

Freiberg, D., Rogers, J., & Serafeim, G. (2019). *How ESG issues become financially material to corporations and their investors*.

Galpin, T., Whittington, J. L., & Bell, G. (2015). Is your sustainability strategy sustainable? Creating a culture of sustainability. *Corporate Governance (Bingley), 15*(1), 1–17. https://doi.org/10.1108/CG-01-2013-0004

Garst, J., Maas, K., & Suijs, J. (2022). Materiality assessment is an art, not a science: Selecting ESG topics for sustainability reports. *California Management Review, 65*(1), 64–90. https://doi.org/10.1177/00081256221120692

Giannessi, E. (1960). *Le aziende di produzione originaria*. Columbo Corsi Editore.

Größler, A., Thun, J. H., & Milling, P. M. (2008). System dynamics as a structural theory in operations management. *Production and Operations Management, 17*(3), 373–384. https://doi.org/10.3401/poms.1080.0023

Hoffman, A. J., & Woody, J. G. (2008). *Climate change: What’s your business strategy?* Harvard Business Press.

Homer, J. (1996). Why we iterate: Scientific modeling in theory and practice. *System Dynamics Review, 12*(1), 1–19. https://doi.org/10.1002/(sici)1099-1727(199621)12:1<1::aid-sdr93>3.0.co;2-p

Homer, J. (2019). Best practices in system dynamics modeling, revisited: A practitioner’s view. *System Dynamics Review, 35*(2), 177–181. https://doi.org/10.1002/sdr.1630

Homer, J., & Oliva, R. (2001). Maps and models in system dynamics: A response to Coyle. *System Dynamics Review, 17*(4), 347–355. https://doi.org/10.1002/sdr.224

IFRS. (2023). *IFRS S1 general requirements for disclosure of sustainability-related financial information*.

Johnson, G., Whittington, R., Scholes, K., Angwin, D., Regnér, P., & Paci, A. (2017). *Strategia. Orientare organizzazioni e imprese in un mondo che cambia* (11th ed.). Pearson Italia.

Jones, E. (2019). Rethinking greenwashing: Corporate discourse, unethical practice, and the unmet potential of ethical consumerism. *Sociological Perspectives, 62*(5), 728–754. https://doi.org/10.1177/0731121419849095

Kaplan, R. S., & Norton, D. P. (1992). The balanced scorecard: Measures that drive performance. *Harvard Business Review., 70*, 71.

Kianto, A., Ritala, P., Spender, J.-C., & Vanhala, M. (2014). The interaction of intellectual capital assets and knowledge management practices in organizational value creation. *Journal of Intellectual Capital, 15*(3), 362–375. https://doi.org/10.1108/JIC-05-2014-0059

Kramer, R. M. (1999). Trust and distrust in organizations: Emerging perspectives, enduring questions. *Annual Review of Psychology, 50*, 569–598.

Kunc, M., Giorgino, M. C., & Barnabè, F. (2021). Developing forward-looking orientation in integrated reporting. *Meditari Accountancy Research, 29*(4), 823–850. https://doi.org/10.1108/MEDAR-12-2019-0664

La Torre, M., Sabelfeld, S., Blomkvist, M., & Dumay, J. (2020). Rebuilding trust: Sustainability and non-financial reporting and the European Union regulation. *Meditari Accountancy Research, 28*(5), 701–725. https://doi.org/10.1108/MEDAR-06-2020-0914

Lazarus, R. S., & Folkman, S. (1984). *Stress, appraisal, and coping*. Springer Publishing Company.

Lewicki, R. J., & Bunker, B. B. (1996). Developing and maintaining Trust in Work Relationships. In R. M. Kramer & T. R. Tyler (Eds.), *Trust in Organizations: Frontiers of theory and research* (pp. 114–139). Sage Publications.

Locke, E. A., & Latham, G. P. (1990). *A theory of goal setting and task performance*. Prentice-Hall.

Maniora, J. (2018). Mismanagement of sustainability: What business strategy makes the difference? Empirical evidence from the USA. *Journal of Business Ethics, 152*(4), 931–947. https://doi.org/10.1007/s10551-018-3819-0

Mclucas, A. C. (2003). *Incorporating soft variables into system dynamics models: A suggested method and basis for ongoing research*.

Meadows, D. H. (2008). *Thinking in systems* (sustainability institute, Ed.).

Millar, E., & Searcy, C. (2020). The presence of citizen science in sustainability reporting. *Sustainability Accounting, Management and Policy Journal, 11*(1), 31–64. https://doi.org/10.1108/SAMPJ-01-2019-0006

Morecroft, J. D. W. (1983). System dynamics: Portraying bounded rationality. *Omega, 11*(2), 131–142. https://doi.org/10.1016/0305-0483(83)90002-6

Morecroft, J. D. W. (1985). Rationality in the analysis of behavioral simulation models. *Management Science, 31*(7), 900–916.

Morecroft, J. D. W. (2015). *Strategic modelling and business dynamics. A feedback systems approach* (2nd ed.). Wiley.

Muñoz-Torres, M. J., Fernández-Izquierdo, M. Á., Rivera-Lirio, J. M., & Escrig-Olmedo, E. (2019). Can environmental, social, and governance rating agencies favor business models that promote a more sustainable development? *Corporate Social Responsibility and Environmental Management, 26*(2), 439–452. https://doi.org/10.1002/csr.1695

OIBR. (2022). *Operationalising materiality–applied guidelines on how to identify and monitor the evolution of sustainability-related material issues*.

Oladimeji, O. O., Keathley-Herring, H., & Cross, J. A. (2020). System dynamics applications in performance measurement research. *International Journal of Productivity and Performance Management, 69*(7), 1541–1578. https://doi.org/10.1108/IJPPM-12-2018-0453

Onida, P. (1953). La dimensione e struttura delle aziende e la produttività. *Produttività, 12*.

Orlitzky, M., Schmidt, F. L., & Rynes, S. L. (2003). Corporate social and financial performance: A meta-analysis. *Organization Studies, 24*(3), 403–441. www.sagepublications.com

Porter, M. E., & Kramer, M. R. (2006). Strategy and society: The link between competitive advantage and corporate social responsibility. *Harvard Business Review*.

Randers, J. (1980). Guidelines for model conceptualization. In J. Randers (Ed.), *Elements of the system dynamics method* (pp. 117–139). The MIT Press.

Randers, J. (1997). Elements of the system dynamics method. *Journal of the Operational Research Society, 48*(11), 1144–1145. https://doi.org/10.1057/palgrave.jors.2600456

Richardson, G. P. (1986). Problems with causal-loop diagrams. *System Dynamics Review, 2*(2), 158–170. https://doi.org/10.1002/sdr.4260020207

Richardson, G. P. (1995). Loop polarity, loop dominance, and the concept of dominant polarity. *System Dynamics Review, 11*(1), 67–88. https://doi.org/10.1002/sdr.4260110106

Richardson, G. P. (1997). Problems in causal loop diagrams revisited. *System Dynamics Review, 13*(3), 247–252. https://doi.org/10.1002/(SICI)1099-1727(199723)13:3<247::AID-SDR128>3.0.CO;2-9

Richardson, G. P. (1999). Reflections for the future of system dynamics. *The Journal of the Operational Research Society, 50*(4), 440. https://doi.org/10.2307/3010465

Richardson, G. P., & Pugh, A. L. (1997). Introduction to system dynamics modeling with DYNAMO. *Journal of the Operational Research Society, 48*(11), 1146–1146. https://doi.org/10.1057/palgrave.jors.2600961

Sachs, J. (2012). From millennium development goals to sustainable development goals. *Lancet, 379*, 2206–2211.

Saeed, K. (1992). Slicing a complex problem for system dynamics modeling. *System Dynamics Review, 8*(3), 251–261.

Schiehll, E., & Kolahgar, S. (2021). Financial materiality in the informativeness of sustainability reporting. *Business Strategy and the Environment, 30*(2), 840–855. https://doi.org/10.1002/bse.2657

Schwaninger, M. (2009). The role of system dynamics within the systems movement. In Y. Barlas (Ed.), *System dynamics: Systemic feedback modeling for policy analysis* (Vol. 1, pp. 266–284). Eolss Publishers Company Limited.

Sterman, J. D. (2000). *Business dynamics: Systems thinking and modeling for a complex world.* Irwin McGraw-Hill.

Stoughton, A. M., & Ludema, J. (2012). The driving forces of sustainability. *Journal of Organizational Change Management, 25*(4), 501–517. https://doi.org/10.1108/09534811211239191

Torelli, R., Balluchi, F., & Furlotti, K. (2020a). The materiality assessment and stakeholder engagement: A content analysis of sustainability reports. *Corporate Social Responsibility and Environmental Management, 27*(2), 470–484. https://doi.org/10.1002/csr.1813

Torelli, R., Balluchi, F., & Lazzini, A. (2020b). Greenwashing and environmental communication: Effects on stakeholders’ perceptions. *Business Strategy and the Environment, 29*(2), 407–421. https://doi.org/10.1002/bse.2373

United Nations. (2023). *Global sustainable development report. Times of crisis, times of change. Science for accelerating transformations to sustainable development.*

van der Waal, J. W. H., & Thijssens, T. (2020). Corporate involvement in sustainable development goals: Exploring the territory. *Journal of Cleaner Production, 252*. https://doi.org/10.1016/j.jclepro.2019.119625

Vignieri, V. (2022). Enhancing performance regimes to enable outcome-based policy analysis in Cross-boundary settings (Vol. 6). Springer International Publishing. doi:https://doi.org/10.1007/978-3-031-07074-7.

Von Bertalanffy, L. (1968). *General system theory: Foundations, development, Applications.* George Braziller.

Waddock, S. A., & Graves, S. B. (1997). The corporate social performance-financial performance link. *Strategic Management Journal, 18*(4), 303–319. https://doi.org/10.1002/(SICI)1097-0266(199704)18:4<303::AID-SMJ869>3.0.CO;2-G

Wiener, N. (1948). *Cybernetics, or control and communication in the animal and the machine.* The MIT Press.

Winch, G. W., Arthur, D. J. W., & Grey, C. (1999). Building and conceptualising generic model structures to address fundamental business change. In *Proceedings of the 18th.* System Dynamics Conference.

Winch, G. W., & Arthur, D. J. W. (2002). User-parameterised generic models: A solution to the conundrum of modelling access for SMEs? *System Dynamics Review, 18*(3), 339–357. https://doi.org/10.1002/sdr.252

Wolstenholme, E. F., & Coyle, R. G. (1983). The development of system dynamics as a methodology for system description and qualitative analysis. *The Journal of the Operational Research Society, 34*(7), 569–581. https://about.jstor.org/terms

Yusliza, M. Y., Yong, J. Y., Tanveer, M. I., Ramayah, T., Noor Faezah, J., & Muhammad, Z. (2020). A structural model of the impact of green intellectual capital on sustainable performance. *Journal of Cleaner Production, 249*. https://doi.org/10.1016/j.jclepro.2019.119334

Chapter 4
Revisiting Sustainability Reporting Through System Dynamics

Contents

4.1 Base Scenario Dynamics

4.1.1 Model Overview

This chapter begins with the analysis of the system behaviour in its base scenario: an idealised situation in which companies respond to increasing external pressures linked to sustainability through reactive behaviours. These responses often aim to

R. Santoni, *The Dynamics of Materiality*, System Dynamics for Performance Management & Governance 8, https://doi.org/10.1007/978-3-032-19280-6_4

preserve existing "business-as-usual" strategies, sometimes relying on opportunistic practices designed to meet external expectations without a substantial revision of internal routines or the business model.

Before presenting the simulation outcomes—expressed through numbers and graphs—it is important to recall the broader meaning and role of the model, briefly summarising the premises developed in the first three chapters of the book.

The System Dynamics model developed here offers a generic structure to simulate, in an aggregated way, both the internal and external dimensions of a company's commitment to a sustainability issue and its performance reporting process. As a generic structure, its purpose is not to provide precise forecasts of specific variables, but rather to offer an interpretive lens to understand the systemic logics that emerge when sustainability is treated as a strategic leverage instead of a compliance burden.

The model simulates a time span from 2015 to 2050, during which sustainability concerns are assumed to grow steadily, reflecting the increasing public relevance of environmental and societal issues. This trend generates a progressive, exogenous rise in stakeholders' expectations, understood as an aggregate of reputational, regulatory, and market pressures that companies are called to manage. By simulating organisational responses to these external demands, the model allows us to observe the long-term repercussions of alternative business choices.

To facilitate the interpretation of the simulation results, three core variables have been selected to represent the essential dimensions of the system behaviour.

The first is *stakeholders' satisfaction*, a synthetic indicator that captures the convergence of external pressures, actual performance, and reputational trust into a single measure of strategic success. This relative value, ranging from 0 to 1, does not necessarily reflect the levels of performance. Rather, it may represent the ability of a company to preserve a strong initial reputation over time, even in the presence of limited improvements.

The second variable is *achieved performance*, which offers an objective assessment of the company's sustainable performance. Also expressed on a scale from 0 to 1, this measure reflects the degree to which the organisation approaches a hypothetical maximum, ideally representing a complete sustainable transition. In scenarios characterised by low external pressure, companies may lack incentives to improve, leading to stagnation and stable performance below the maximum level of 1.

The third key dimension is *financial achievements*, a metric that integrates both the relevance of the sustainability issue and the level of *stakeholders' satisfaction*. This variable ranges from 0 to 4, based on the model's calibration parameters (see Table 3.3 in Sect. 3.3.6), and reflects the financial returns linked to strategic alignment with sustainability concerns.

Beyond these three indicators, the model includes a set of additional variables that act as drivers—either positive or negative—of the system's evolution. These include: *stakeholders' expected performance*, *communicated performance*, *greenwashing performance communication*, *perceived greenwashing*, *relative greenwashing*, *sustainability-dedicated assets and competencies*, *internal materiality*, *stakeholders' materiality*, and *average global materiality*. These elements are visualised in the graphs throughout this chapter, offering a more detailed understanding of the dynamics at play during the simulation runs.

With the model structure and key variables now introduced, we turn to the outcomes of the base scenario simulation. Section 4.1.2 presents the results of this initial run, which serves as a reference trajectory for evaluating alternative strategies and identifying potential leverage points.

4.1.2 *Simulation of the Initial Model Settings: The "Base Run"*

The first simulation scenario, named "Base run", serves as a benchmark to explore how the system behaves when companies initially respond to growing sustainability pressures with minimal strategic adaptation, trying to preserve existing routines and business models (i.e., business-as-usual).

Among the model's initial settings, the simulation adopts a double materiality approach, which attributes equal importance to both the financial and impact dimensions of sustainability.

The scenario is the same for all simulations: The *average global materiality* level grows steadily over time, holding an initial relatively low level of public significance (i.e., value of the variable = 0.1 in 2015) and reaching full relevance by 2050 (i.e., value of the variable = 1). This rising materiality acts as an exogenous driver, shaping stakeholders' expectations and influencing organisational responses.

The results of the "Base run" are presented progressively, highlighting the feedback loops that govern the system behaviour across different phases. Figures 4.1, 4.2, and 4.3 break down the simulation into distinct time periods, each characterised by specific dynamics.

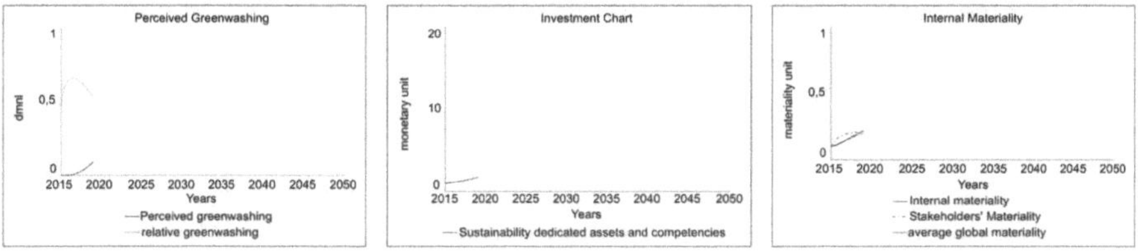

Fig. 4.1 Base Run simulation—years 2015–2019. Source: Author's elaboration

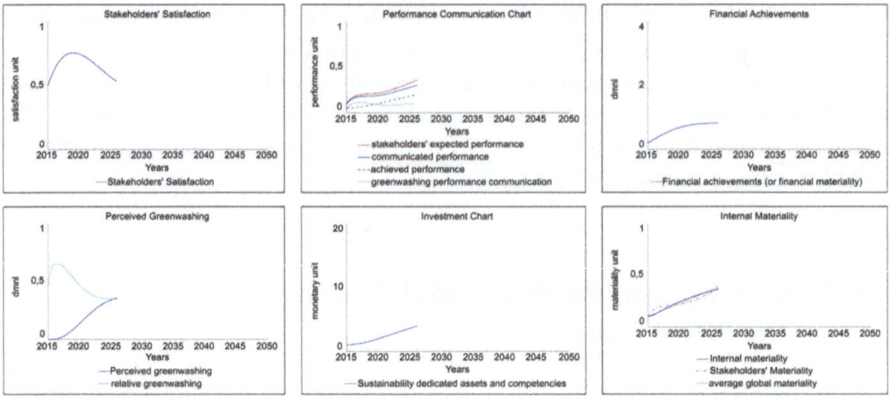

Fig. 4.2 Base Run simulation—years 2019–2026. Source: Author's elaboration

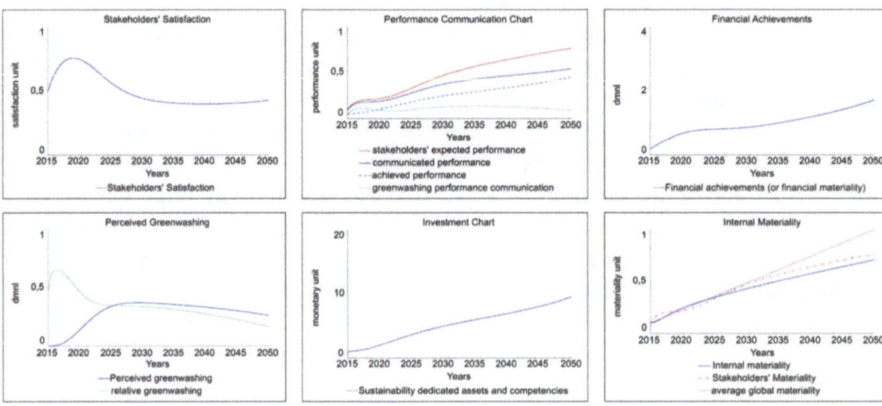

Fig. 4.3 Base Run simulation complete results. Source: Author's elaboration

Each figure follows a consistent layout. The top three graphs display the main outcome indicators:

- *Stakeholders' satisfaction*, capturing the perceived legitimacy of the company
- *Performance communication*, comparing stakeholder *expectations* (red) to *communicated performance* (solid blue), also articulated into its two components: *achieved performance* (dashed blue) and *greenwashing performance communication* (dotted green)
- *Financial achievements*, reflecting the economic returns derived from the interplay between performance and issue relevance

The bottom three graphs provide insight into the underlying mechanisms:

- Greenwashing and scepticism, represented by *relative* (dotted green) and *perceived greenwashing* (blue)
- Investment levels, measured through *sustainability-dedicated assets and competencies*
- Materiality perspectives, including *internal materiality* (blue), *stakeholders' materiality* (red-dashed), and *average global materiality* (dotted pink)

This visual structure allows the reader to trace the evolution of the system over time, identifying the dominant feedback loops and the related consequences on the system behaviour.

During the first years of the simulation (years 2015–2019), illustrated in Fig. 4.1, the steady rise in *average global materiality* activates the reinforcing loop *R1: stakeholders' pressure process*. As sustainability concerns become more prominent, stakeholders begin to raise their expectations (*stakeholders' expectations*) regarding the company's commitment and transparency.

Despite these dynamics, the company's response remains limited: Sustainability is still perceived as an external issue, lacking strategic relevance, as reflected by low levels of *internal materiality*. This perception holds managerial significance: *Investment levels* remain low, and *achieved performance* is far from *stakeholders' expectations*.

In this context, greenwashing communication is considered a valid solution and is adopted as a tactical response to external pressure, neglecting its long-term risks. This stimulates a compensating mechanism, exemplified by the *B2: greenwashing communication immediate effect*, which aims to satisfy stakeholder needs through high levels of *greenwashing performance communication*.

At first, stakeholders believe the company's *communicated performance* and their *satisfaction* increase rapidly. This growing satisfaction translates into improved *financial achievements*. As a result, the company begins to perceive sustainability as a business-relevant issue (driven by a modest rise in *internal materiality*) and gradually reduces the misleading communication practices.

However, beneath the surface, scepticism begins to grow, marking the emergence of the *R4: greenwashing trap*, a reinforcing loop that gradually erodes stakeholder trust and reveals the fragility of superficial sustainability strategies.

During the following years of the simulation (2019–2026), illustrated in Fig. 4.2, the system enters a phase of instability.

Stakeholders' satisfaction drops sharply, driven by growing scepticism that undermines trust in the company's *communicated performance*. This shift exemplifies the generalised sense of dissatisfaction that the population is experiencing nowadays, when the attention to sustainability is higher than in the past, but also the business communications are known as often unfaithful and misleading.

During this period, *perceived greenwashing* converges with *relative greenwashing*, although their trajectories diverge: The former continues to rise, while the latter begins to decline. This divergence signals a growing mismatch between what the company communicates and how stakeholders interpret it.

These conflicting incentives activate opposing feedback mechanisms, generating a form of unstable equilibrium.

On one hand, the company attempts to increase *investments* and *achieved performance* and reduce *greenwashing*, following what should be the natural path of sustainability transition—represented by the balancing loop *B1: company's answer to pressure*. On the other hand, uncertainty prevails. *Financial achievements* stagnate and external pressure intensifies, as shown by rising *stakeholders' materiality*.

These dynamics are symptomatic of the *R4: greenwashing trap*, a reinforcing loop that supports greenwashing rather than sustainable investments, due to a diffused sense of external distrust in the company's communications. The company finds itself caught between the intention to improve and the fear that its efforts will not be recognised, leading to hesitation and strategic inertia.

The final phase of the simulation (2026–2050), illustrated in Fig. 4.3, confirms the system's tendency to settle into a state of imperfect equilibrium. *Stakeholders' satisfaction* stabilises at an intermediate level, reflecting a partial recognition of the company's efforts, but also a persistent lack of full legitimacy. The sustainability transition is thus not achieved.

Going more into details, the company acknowledges its limitations in becoming a sustainability leader, yet accepts a degree of responsibility under external pressure. Misleading communication is reduced, and the gap between *stakeholders' expected* and *communicated performance* is no longer covered, resulting in what could be interpreted as an imperfect but more honest sustainability report.

From a System Dynamics perspective, this scenario is shaped by the interaction of feedback loops. The *R1: stakeholders' pressure process* is partially balanced by investments and performance through *B1: company's answer to pressure*. However, two potentially transformative loops—*R2: financial materiality cycle* and *R3: available resources cycle*—remain inactive. Their dormancy is a long-term consequence of the *R4: greenwashing trap*, which continues to fuel scepticism and inhibit trust until the end of the simulation.

As can be seen in the Greenwashing and Scepticism graph in Fig. 4.3, the trajectory of the variable *perceived greenwashing* reproduces the reference mode of behaviour hypothesised in Figure 3.2 (Sect. 3.2.2). After an initial period marked by intense greenwashing activities, stakeholder scepticism grows and stabilises over time. In this way, the model confirms its ability to replicate the systemic logic whereby greenwashing, once perceived, generates persistent distrust, a dynamic that is difficult to reverse, even when communication practices improve.

However, unlike the hypothesised decline in scepticism at the end of the simulation (see Figure 3.2), the "Base run" shows maintained levels of *perceived greenwashing*. This divergence is due to the absence of external policies or corrective measures aimed at disincentivising greenwashing. The model shows that, by relying solely on internal self-balancing mechanisms, restoring trust once it has been compromised becomes a highly challenging task that can undermine the entire sustainability transition.

Section 4.2 will investigate whether and how this inertia can be altered. Through a sensitivity analysis of key model parameters and strategic variables, the aim is to identify leverage points capable of overcoming undesirable systemic behaviours and promoting a more effective sustainability transition.

4.2 The Results of the Sensitivity Analysis

Section 4.1 will illustrate the dynamics of the base scenario, showing how limited strategic integration of sustainability can foster greenwashing practices and scepticism, which may hinder the transition and lead to systemic stagnation. To explore whether and how this trajectory can be altered, this section presents the results of a sensitivity analysis conducted on the System Dynamics model.[1]

Sensitivity analysis serves two main purposes. First, it tests the robustness of the model by simulating how alternative settings of key parameters affect the system behaviour. Second, it identifies potential leverage points—areas where the model shows sensitivity to parametric changes—signalling potential spaces for managerial intervention.

The results are summarised in Table 4.1, which reports the parameter name, its sector, the base value, the tested range, and the type of sensitivity observed.[2]

Three types of sensitivity are distinguished:

1) Numerical sensitivity: Changes in assumptions produce only minor variations in results. These cases are generally not discussed further, except when they concern variables that were expected to have a significant impact but do not (e.g., weight of financial materiality).
2) Behavioural sensitivity: Changes in assumptions lead to significant shifts in system behaviour, but these parameters are outside the space of managerial intervention. They mainly relate to modelling choices and assumptions, and for this reason, some are treated in the "Appendix A".
3) Policy sensitivity: Parameters that have high impact on the system behaviour and that managers can influence directly. This suggests the existence of potential leverage points for strategic action that will be discussed in this section.

The results of the sensitivity analysis reveal some interesting insights.

The first sector on internal materiality does not show any sensitivity beyond the numerical one. This is true for both the *effects* of the *financial* and *impact* components, as well as to the *initial* levels of *internal materiality* and the increase or

[1] For reproducibility purposes, the analysis was performed using Sobol Sequence Sampling (limited runs: 100 runs) on Stella Architect software, adopting a uniform distribution for parameter variation. The tested ranges were defined to preserve the real-world meaning of each parameter.

[2] Some of the variables listed in Table 4.1 were not shown in the previous stock-and-flow diagrams, as those diagrams illustrated only the main variables. Many of these additional variables result from sub-components necessary for the development of the model. The sensitivity analysis refers to these more detailed variables; however, readers can find a comprehensive description of each parameter and its meaning in the model documentation provided in "Appendix B".

Table 4.1 Results of the sensitivity analysis

Sector	Parameter	Base value	Tested range	Sensitivity
Sector 1: Internal materiality*	Effects of financial on internal materiality (graphical)	Decreasingly increasing	Linear, decreasing, s-shape, increasingly increasing	Numerical
	Initial internal materiality	0.1	0; 0.5	Numerical
	Reference internal materiality	0.5	0.4; 0.6	Numerical
	Sensitivity of internal materiality to impact materiality	1	0.8; 1.2	Numerical
	Time to adjust internal materiality	2	1; 3	Numerical
	Weight of financial materiality	0.5	0.25; 0.75	Numerical
Sector 2: Resource allocation	Asset depletion time	10	8; 15	Numerical
	Asset and competencies Per financial achievement unit	1	0.75; 1.25	Numerical
	Initial available resources	10	5; 15	Policy*
	Initial sustainability assets and competencies	1	0; 5	Behavioural*
	Investment time	3	2; 4	Numerical
	Greenwashing propensity	0.5	0; 1	Policy*
	Normal percentage of resources	0.25	0; 1	Policy*
	Resource allocation sensitivity to internal materiality	1	0.5; 1	Numerical
	Sensitivity of greenwashing to internal materiality	−1	−0.5; −1	Numerical
	Sensitivity of reinvestments to materiality	0.5	0.25; 0.75	Numerical
	Threshold of internal materiality for resource allocation	0.5	0.3; 0.7	Policy*

(continued)

Table 4.1 (continued)

Sector	Parameter	Base value	Tested range	Sensitivity
Sector 3: Performance communication	Effect of investments on performance (graphical)	Decreasingly increasing	Linear, increasingly increasing, s-shape	Numerical
	Expectation sensitivity to materiality	1	0.75; 1	Numerical
	Initial stakeholders' satisfaction	0.5	0; 1	Behavioural**
	Normal dedicated investments	5	3; 7	Behavioural**
	Normal expected performance	0.5	0.4; 0.6	Numerical
	Normal performance level	0.25	0.125; 0.375	Behavioural**
	Perceived greenwashing	SMTH 3	SMTH 1;3;5	Numerical
	Perceived greenwashing adjustment time	9	5; 15	Numerical
	Perceived greenwashing multiplier on communication	Decreasingly decreasing	Linear, decreasingly decreasing, z-shape, slowly decreasing, different mins and max	Behavioural**
	Time to adjust satisfaction	3	1; 5	Numerical
Sector 4: Impact materiality	Average global materiality (graphical)	Linearly increasing	Increasingly increasing, s-shape	Numerical
			Constant 1	Behavioural**
	Reference stakeholders' materiality	0.5	0.4; 0.6	Numerical
	Satisfaction weight	0.5	0; 1	Numerical
	Time to adjust stakeholders' materiality	2	1; 5	Numerical
Sector 5: Financial materiality	Maximum market dimension	4	3; 5	Numerical
	Sensitivity of financial achievements to satisfaction	0.5	0.4; 0.6	Numerical
	Market dimension sensitivity to stakeholders' materiality	0.5	0.4; 0.6	Numerical

Key: "*" means further explanation is provided in this chapter; "**" means further explanation is provided in "Appendix A"
Source: Author's elaboration

decrease of 1 year in the frequency of the materiality assessment. Interestingly, no significant sensitivity is detected even regarding the choice of an impact, double, or financial-oriented materiality approach, represented by the variable *weight of financial materiality*. This absence of sensitivity will be further explored in Subsection 4.3.1.

A different picture emerges for Sector 2, which is the part of the model related to resource allocation. This is the area of greatest managerial intervention and encompasses all the spaces of policy sensitivity within the model, which emerged in four cases.

The first two concern the parameters *initial available resources* and *normal percentage of resources*. The former indicates the quantity of resources on which the company can count all over the simulation to increase sustainability investments. As could be expected, the availability of financial resources contributes to the potential of a company to attain sustainability transition. The second parameter concerns the speed at which these *available resources* can be invested. For this reason, these two sensitivities are discussed together in Subsection 4.3.2.

The third case concerns the parameter *greenwashing propensity*: Changing the company's attitude towards misleading communication of performance has a significant impact on the long-term simulation outputs. Alternative scenarios are presented in Subsection 4.3.3.

Finally, the fourth policy sensitivity is represented by the *threshold of internal materiality for resource allocation*. This *threshold* describes the level of materiality required to invest in sustainability. Here too, different values influence the results, as shown in Sect. 4.3.4.

Sector 2 also presents a case of behavioural sensitivity for the variable *initial sustainability assets and competencies*. This value is considered non-influenceable by managers, as it represents an initial state of the corporate system that cannot be retroactively modified: Managers can only decide how many resources to invest in the future, while a different initial investment level represents only a different corporate scenario. In this sense, this discussion differs from that regarding available resources, since in that case, managers could allocate more existing resources or draw on new sources of capital for future investments.

Finally, the remaining variables show only numerical sensitivity. This demonstrates the robustness of the model: Even when using different formulations and weights, the system would exhibit similar behaviours throughout the simulations.

Moving to the third sector on performance communication, this is the one that shows the greatest behavioural sensitivity to the initial settings of variables and parameters. However, no space for managerial initiatives is identified, and the sensitivity is consistent with the model's formulation. In detail:

- The *initial* level of *stakeholders' satisfaction* influences the entire outcome of the simulation. This is intuitive: A company that can count on excellent relationships with its stakeholders is extremely advantaged compared to a company operating in a context of general dissatisfaction. Consequently, it will face less pressure, achieve higher economic returns, and have less incentive to greenwash, and its efforts can just focus on maintaining the trust and satisfaction gained in previous years.

- The variables *normal dedicated investments* and *normal performance level* indicate, respectively, the level of investments required to reach the performance level considered normal, and the same normal performance level. This sensitivity is explained by the model's formulation: If achieving the same level of performance requires more investments, it is obvious that the transition will take longer and the overall simulation will be worse.
- Changing the graphical function of the impact of scepticism on *stakeholders' satisfaction* (*perceived greenwashing multiplier on communication*) affects the entire system behaviour. The original formulation indicates that, as *perceived greenwashing* increases, the credibility of communication decreases more than proportionally, since minor trust violations have a higher marginal impact in the beginning, creating doubt and uncertainty (Lewicki & Bunker, 1996; Slovic, 1993). Removing this assumption modifies simulation results, but no manager can act on this lever.
- As for the formulation of *perceived greenwashing*, numerical sensitivity indicates that no significant differences are observed when varying the time settings and delay order.

Finally, the discussion of the fourth and fifth sectors can be addressed jointly. Here, except a single case, all variables show only numerical sensitivity. This confirms the model's robustness even in a simplified formulation such as that of the financial sector.

The only case of behavioural sensitivity concerns the external driver of the model: *average global materiality*. First of all, it should be noted that the way in which materiality grows (constantly, more slowly, or more rapidly) does not significantly change the system's response or the simulation outcomes. A different situation arises when the level of *average global materiality* of a sustainability issue starts at its maximum value from the very beginning and remains stable throughout the simulation. This scenario simulates an already ongoing emergency, where the issue is at the highest level of relevance from the start. In this case, the transition occurs much more rapidly: The business-as-usual model is quickly abandoned, and the greenwashing alternative is not considered. This is a highly relevant finding, as it explains that if the sense of urgency were truly perceived by the entire population, the system would be capable of achieving a rapid and comprehensive transition.

In conclusion, the sensitivity analysis confirms the robustness of the model while highlighting a limited set of parameters with significant policy sensitivity. Building on these insights, the next Sect. 4.3 explores alternative scenarios designed to test the practical implications of these leverage (or non-leverage) points.

4.3 Simulating Alternative Scenarios

Building on the insights from the sensitivity analysis, this section explores how specific strategic choices can reshape the system behaviour and, in some cases, accelerate the sustainability transition. In particular:

- 4.3.1—Simulating alternative materiality approaches: Examines how a financial, impact, or double approach to materiality assessment can influence system dynamics
- 4.3.2—Simulating the effects of the resource allocation lever: Assesses the effects of increasing initial resources and accelerating investment flows
- 4.3.3—Simulating alternative greenwashing propensities: Evaluates the consequences of increasing or reducing the adoption of misleading communication practices
- 4.3.4—Simulating different levels of the internal materiality threshold: Analyses how changes in the internal materiality threshold can alter investments and reliance on greenwashing.

These simulations allow to move beyond the identification of theoretical leverage points to determine whether and how alternative configurations can overcome undesirable behaviours observed in the "Base run", thereby setting the stage for the discussion in Sect. 4.4.

4.3.1 Simulating Alternative Materiality Approaches

After having reviewed the evolution of the concept of sustainability in business—from its origins to the debate on materiality as the last major source of heterogeneity in the standardisation of reporting guidelines—it is legitimate to ask how alternative approaches influence corporate behaviour. In particular, how would the results of the "Base run", which adopts a double materiality approach, change if a company chose an approach more oriented towards financial or impact materiality?

In the model, this aspect is represented by the variable *weight of financial materiality*, which determines the relative importance of financial information in the materiality assessment process. The sensitivity analysis revealed that changes in this parameter, tested within the range 0.25–0.75,[3] produced only numerical sensitivity. In other words, the system's overall behaviour remains largely unchanged,

[3] The range 0.25–0.75 was selected according to the following reasoning:

- A value of 0.5 attributes equal importance to financial and impact dimensions, recreating a double materiality approach.
- A value of 0.25 simulates the behaviour of an impact-oriented company. For the model structure, attributing a value of 0 to the variable would have meant neglecting the financial results, obstructing transition due to financial resource scarcity.

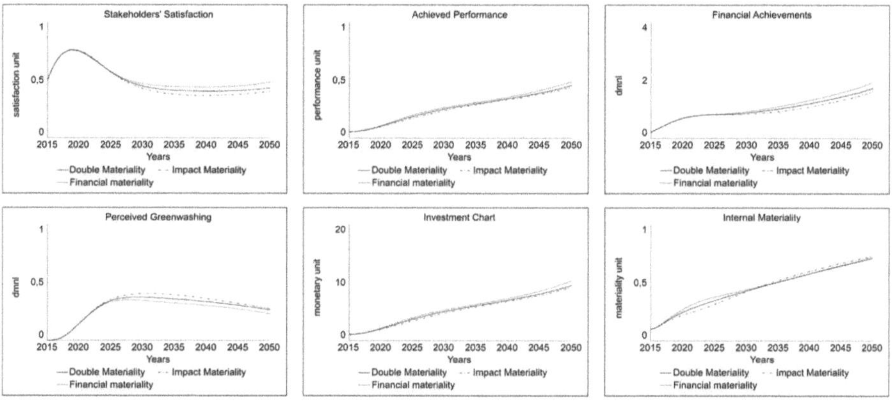

Fig. 4.4 Simulations of alternative materiality approaches. Source: Author's elaboration

regardless of whether the company adopts a financial, impact, or double materiality approach.

Figure 4.4 illustrates this comparison: The blue line represents the "double materiality" approach (weight = 0.5), the red-dashed line the "impact materiality" approach (weight = 0.25), and the pink-dotted line the "financial materiality" approach (weight = 0.75). While the financial orientation appears slightly dominant, the differences are minimal and largely attributable to model formulation, as explained below.

Impact-oriented companies tend to align their behaviour with *stakeholders' materiality*, while financially oriented firms respond primarily to market opportunities. In the presence of greenwashing, which characterises the initial phase of all simulations, both orientations initially benefit from the increase in *stakeholders' satisfaction*, which reduces pressure and postpones the urgency to invest. Financially oriented companies interpret the increase in revenues associated with satisfaction as an incentive to invest, whereas impact-oriented firms delay action until external pressure intensifies. These dynamics, however, do not substantially alter the system's trajectory.

To isolate the effect of materiality orientation, additional simulations were performed under conditions where greenwashing was removed—that is, *normal greenwashing* set to zero. While the substantial differences between scenarios with or without greenwashing will be explained in Sect. 4.3.3, the focus here is only on the differences among the three approaches shown in each graph.

Figure 4.5 shows that, in this scenario, the gap between the three approaches becomes more visible in terms of final results, with the impact approach

- A value of 0.75 simulates the approach of a financial-oriented company. In this case, a value of 1 would have meant ignoring the general relevance of a sustainability issue, including its market dimension, in favour of the analysis of the achieved financial results only.

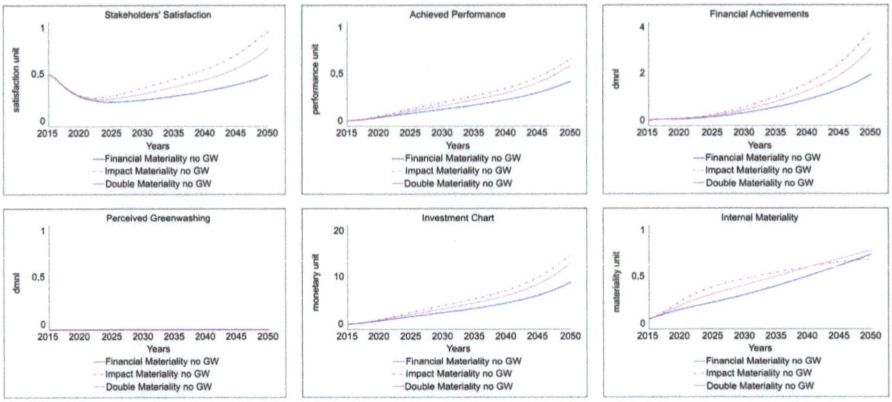

Fig. 4.5 Simulations of alternative materiality approaches, greenwashing excluded. Source: Author's elaboration

outperforming the double and financial ones. This happens because impact-oriented firms invest earlier, without waiting for sustainability issues to gain financial relevance. This allows them to secure a stronger market position if the issue later becomes economically significant. Yet, even in this case, the differences remain primarily numerical: the general trend of the system is maintained regardless of the materiality approach adopted.

It is also possible to test alternative scenario settings to further explore the implications of materiality orientation. For instance, what happens if a sustainability issue gains moral relevance but not market significance? This condition can be simulated by halving the parameter *maximum market dimension* from 4 to 2. Figure 4.6 illustrates this scenario. As expected, impact-oriented companies achieve better results, but once again, the difference remains only numerical. Without financial repercussions, the strategy lacks economic sustainability: Once the stock of *available resources* is depleted, *financial achievements* are insufficient to fuel the transition. In this sense, even the most impact-oriented companies are still businesses and for this require financial resources to survive.

This concludes the tests on materiality approaches, confirming the numerical sensitivity of the model to the way materiality assessment is conducted. The lack of sensitivity in this first sector can be explained by the model's focus on the managerial processes that follow materiality assessment. In practical terms, companies are generally well aware of sustainability challenges; therefore, achieving satisfactory sustainability performance is not a matter of how materiality is assessed, but of how it is integrated into corporate decision-making.

The next subsections explore simulations based on the leverage points identified in the sensitivity analysis, focusing on areas where managerial action can meaningfully influence system behaviour.

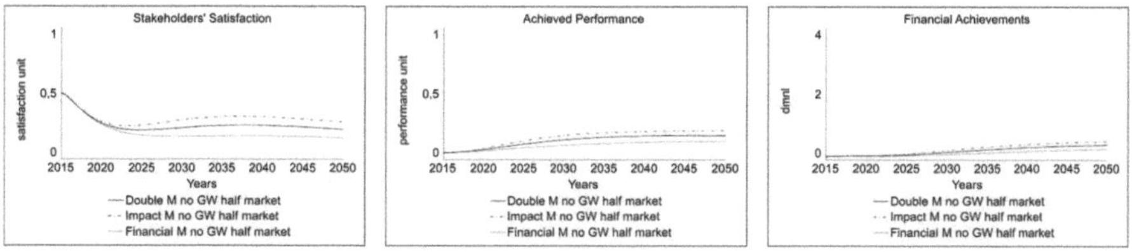

Fig. 4.6 Simulation results: Decreasing the financial relevance of a sustainability issue. Source: Author's elaboration

4.3.2 Simulating the Effects of the Resource Allocation Lever

While the existence of a leverage point in resource investment may seem intuitive, real-world experience shows that even in contexts with abundant resources, the transition often struggles to take off. Indeed, managers tend to be constrained by cost-benefit reasoning or by path dependency linked to business-as-usual practices, postponing investments in sustainability.

The "Base run" captured this dynamic: The available resources at the company's disposal were sufficient to promote the transition, yet the simulation outcomes revealed delayed investments, which became less incisive due to the prolonged stagnation dominated by scepticism.

The results of the sensitivity analysis displayed in Table 4.1 suggest that acting on the resource allocation process could help overcome these dynamics, thanks to two levers:

- *Initial available resources*, which indicate the amount of resources the company makes available for the entire duration of the simulation.
- *Normal resource allocation*, which represents the percentage of resources that can be allocated within a hypothetical 3-year investment period.

These are analysed in turn in the following paragraphs.

4.3.2.1 Changing the Initial Endowment of Available Resources

Model sensitivity to resource endowments was assessed by assigning three different values to the variable *initial available resources*: one corresponding to the "Base run" (10 monetary units, blue line), and two others with a 50% increase and decrease (15 and 5 monetary units, red-dashed and pink-dotted lines, respectively). It is important to note that no other settings were modified: these resources represent only the stock of *available resources*, and the company will autonomously decide whether to keep or invest them.

Figure 4.7 illustrates these dynamics.

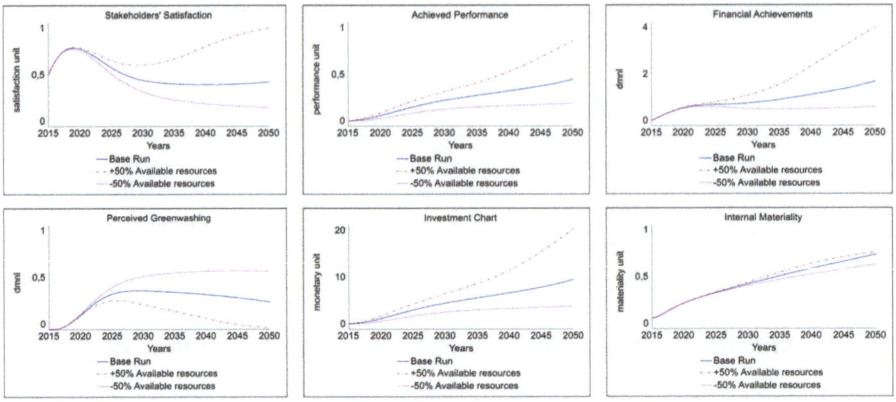

Fig. 4.7 Simulation results: Initial available resources leverage. Source: Author's elaboration

Here, the magnitude of change in presence of policy sensitivity becomes evident. Unsurprisingly, the scenario with more resources achieves higher performance than the others. This occurs across all sectors, as reflected in very high levels of *stakeholders' satisfaction*, *achieved performance* reaching almost the maximum level at the end of the simulation, and *financial achievements* more than doubling those of the "Base run". Even legitimacy, expressed in terms of *perceived greenwashing*, is different: A company with a larger amount of resources to invest can react more effectively to a loss of legitimacy; once scepticism is perceived, the company can more easily abandon misleading practices in favour of real investments in sustainability.

However, the other two underlying graphs show that: (1) The level of *internal materiality* does not show significant differences, signalling that the relevance of the topic is similar for companies with different investment potential; (2) the investment chart displays only a narrow initial gap, which becomes larger towards the end of the simulation, when *financial achievements* are great and fuel reinvestments. This indicates that behavioural differences may not be caused by total resource endowment per se, but by early allocation strategies aimed at gaining competitive advantage. For this reason, the next paragraph focuses on the allocation speed as a further element of leverage.

4.3.2.2 Changing the Resource Allocation Speed

The second parameter characterised by policy sensitivity concerns the speed at which available resources are invested. As mentioned earlier, the parameter *normal resource allocation* expresses the percentage of resources that can be invested within a 3-year period, assuming *internal materiality* equals the *threshold*. Doubling the value of this parameter from 0.5 to 1 means doubling the speed of resource allocation, thereby accelerating the company's transition towards sustainability. It is

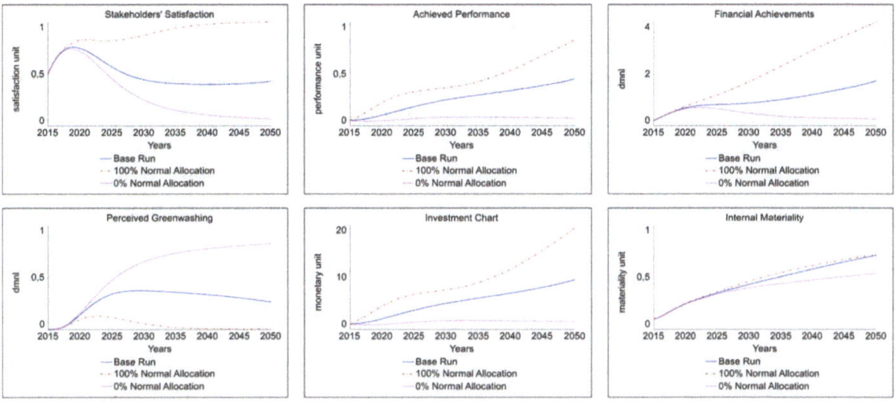

Fig. 4.8 Simulation results: Normal resource allocation leverage. Source: Author's elaboration

worth noting that the amount of *available resources* used in this simulation is the same as in the "Base run", not the increased amount used in Fig. 4.7.

Figure 4.8 compares the three scenarios:

- The "Base run" with original settings (*normal resource allocation* = 0.5, blue line).
- A scenario called "100% normal allocation", where the parameter is doubled (*normal resource allocation* = 1; red-dashed line).
- A scenario, named "0% normal allocation", where the parameter is set to zero, meaning the company cannot rely on any resources other than those generated by *financial achievements* (*normal resource allocation* = 0; pink-dotted line).

Although this parameter acts on the same leverage points as the previous simulation, the results reveal some differences with high managerial significance. The main difference concerns the trajectory of the *stakeholders' satisfaction* curve: Here, the consequences of scepticism are minimal. Indeed, the company provides tangible proofs of its commitment to sustainability by investing resources in a significant and rapid way, fulfilling stakeholder expectations.

Surprisingly, despite fewer resources than in the Fig. 4.7 scenario, the company achieves better results, particularly during the central phase of the simulation. These attainments can be seen in the *financial achievements* curve, which grows linearly, meaning that the company acts as a market leader, achieving the benefits of market expansion with revenues that increase steadily. Moreover, the investment chart shows a wave pattern in which, after an initial increase, investments slow down to collect the benefits of an accelerated transition and resume later in the second half of the simulation.

In this context, the *achieved performance* at the end of the simulation is slightly lower than in the scenario of Fig. 4.7. This illustrates the relative nature of sustainability: A company that demonstrates alignment with sustainability principles over the long term can satisfy stakeholders with lower absolute results compared to one that needs to overcome a perceived legitimacy gap.

All feedback loops contribute to explaining the system behaviour. In both cases, the greater or faster allocation of financial resources to sustainability investments primarily strengthens the balancing loop *B1: company's answer to pressure*, which reduces stakeholder concerns by meeting their informational needs, while simultaneously activating the self-reinforcing loops *R2* and *R3* on financial materiality and newly investable resources.

In the case of rapid investments, the greenwashing and scepticism loops are deactivated early, while high *stakeholders' satisfaction* neutralises the *R1: stakeholders' pressure process*.

While the managerial implications of the resource allocation leverage points will be explored in Sect. 4.4, it is now the moment for moving to another critical policy sensitivity: greenwashing propensity. This shifts the focus from performance prerequisites to the communication domain, addressing the integrity with which companies meet stakeholders' informational needs.

4.3.3 Simulating Alternative Greenwashing Propensities

The sensitivity analysis highlighted *greenwashing propensity* as a critical factor in explaining system behaviour. Unlike parameters related to internal efficiency or resource management, this variable reflects a strategic choice: whether to communicate performance transparently or opportunistically.

The model sensitivity to *greenwashing propensity* is driven by the activation of feedback loops with contrasting polarities. In this analysis, two loops are particularly relevant: *B2: greenwashing communication immediate effect*, which captures the short-term increase in *stakeholders' satisfaction* generated by positive communication, and *R4: greenwashing trap*, which illustrates the long-term repercussions of growing scepticism. As distrust rises, firms may paradoxically resort to further greenwashing, perceiving investments as vain when stakeholders no longer trust reported results.

The results of the simulations carried out by modifying the values of the *greenwashing propensity* parameter are shown in Fig. 4.9. It shows, alongside the "Base run" (blue line), the following scenarios:

- "max greenwashing" (*greenwashing propensity* = 1; red-dashed line)
- "low greenwashing" (*greenwashing propensity* = 0.25; green-dashed line)
- "very low greenwashing" (*greenwashing propensity* = 0.1; pink-dotted line)
- "0 greenwashing" (*greenwashing propensity* = 0; yellow solid line).

While the "max greenwashing" case represents a situation in which the company—for levels of *internal materiality* equal to the *threshold*—compensates the entire gap between *achieved performance* and *stakeholders' expectations* through

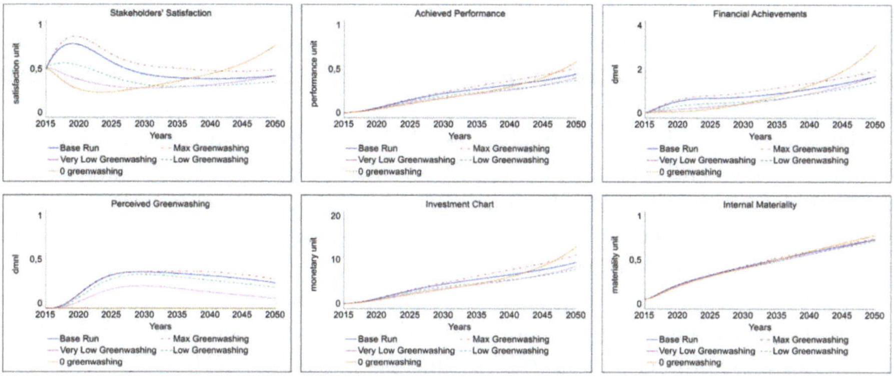

Fig. 4.9 Simulation results: Greenwashing propensity leverage. Source: Author's elaboration

greenwashing mechanisms, the others represent progressively lower compensation, up to the exclusion of any form of greenwashing in the "0 greenwashing" case.

The type of sensitivity displayed by the related graphs differs from what emerged in the previous case of resource allocation. In fact, policy sensitivity appears only for a specific value of the parameter, namely *greenwashing propensity* equal to zero. The main behavioural difference concerns the trend of the curves at the end of the simulation: Unlike the other four cases, the last scenario represents a company in clear affirmation and market launch that can count on a distinct competitive advantage over the other four examples.

However, a company that does not engage in greenwashing is the worst performer of the first part of the simulation. The initial level of investment is inadequate to meet stakeholder needs, resulting in a decrease in satisfaction and increased pressure. Only in the second half of the simulation, thanks to a strategic adjustment in materiality and investments, the company succeeds in entering the sustainability market. This happens in a more decisive way than its competitors, because the company does not suffer the effects of scepticism, with the variable *perceived greenwashing* remaining fixed at zero throughout the simulation.

As concerns the comparison among the four situations that include some form of greenwashing, the "max greenwashing" scenario results clearly dominant. Although this result may seem counterintuitive at this point in the discussion, it actually describes a well-defined situation: If the company intends to adopt misleading communication practices, the choice to do it fully dominates over lower levels of greenwashing. This highlights an important aspect: It is almost pointless to talk about different strategic levels of greenwashing. Greenwashing is done or it is excluded.

To conclude the description of the results, the behaviour of the system in the "0 greenwashing" scenario can be interpreted through feedback loop analysis. Setting *greenwashing propensity* to zero deactivates the loops *R4: greenwashing trap* and *B2: greenwashing communication immediate effect*, reducing the overall model complexity. The initial loop *R1: stakeholders' pressure process* drives a reinforcing increase in *stakeholders' expectations*, which quickly lowers *stakeholders' satisfaction*. This pressure is internalised through *internal materiality*, activating *B1: company's answer to pressure*. At this point, sustainability issues are internalised, and the company invests to improve performance.

The *financial achievements* graph shows that although this situation takes time to emerge, the loops *R2: financial materiality cycle* and *R3: available resources cycle* can foster exponential growth. Their activation marks a turning point: The company begins to view sustainability as a source of success, reinvesting financial gains and reshaping mission, vision, and values.

What emerges is that anticipating the activation of these loops is crucial, as they accelerate growth and the sustainability transition. The next subsection shows how the strategic use of *internal materiality* can support this acceleration.

4.3.4 Simulating Different Levels of the Internal Materiality Threshold

The third and last area of policy sensitivity concerns the *threshold of internal materiality for resource allocation*. As explained earlier (see Sect. 3.2.3.2), *internal materiality* acts as a "switch" capable of diverting business strategy from greenwashing towards genuine sustainability. This switch is controlled by the *threshold*, which determines whether efforts flow into investments or misleading communication.

Mathematically, the *threshold* influences resource allocation positively and greenwashing negatively: A lower *threshold* allows resources to move faster from the available stock into *sustainability-dedicated assets and competencies* and favours the reinvestment of *financial achievements*, reinforcing the company's commitment to sustainability.

The simulation results in Fig. 4.10 compare the "Base run" (*threshold* = 0.5, blue line) with two scenarios where the *threshold* is reduced to 0.4 (red-dashed) and 0.3 (pink-dotted).[4]

The charts highlight the significant potential of this managerial lever. The stagnation trend of *stakeholders' satisfaction* shown in the "Base run" is strongly countered by lowering the *threshold*. Likewise, both *achieved performance* and *financial achievements* are highly sensitive to this adjustment. The main reason lies in the investment chart: the strategic meaning of the *threshold* directs more resources to the stock of *sustainability dedicated assets and competencies*, influencing the system

[4]As shown in the simulations, lower *thresholds* can lead to *achieved performance* exceeding 1 before the end of the run, which may suggest over-investment. However, this is not an error but a consequence of using a fixed threshold in a dynamic context: What is "low" at the beginning becomes too permissive later. For instance, *average global materiality* reaches 0.3 in 2023, 0.4 around 2027, and 0.5 only after 2030. Setting a threshold of 0.3 means being highly proactive in the early years but, by 2050, allows investments far beyond what is strategically necessary. This explains why performance can exceed 1.

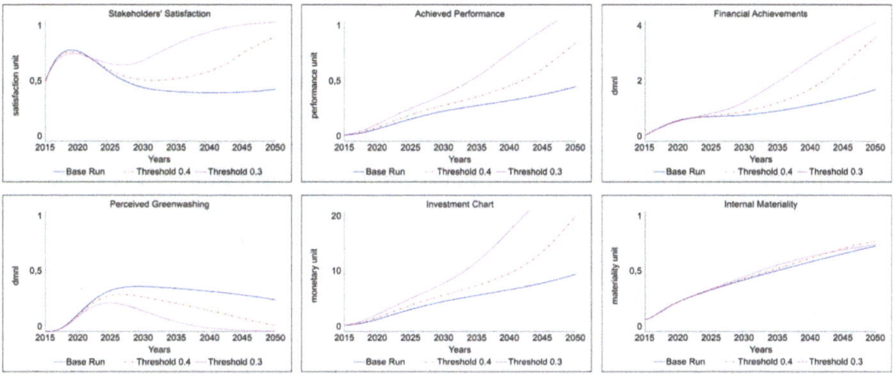

Fig. 4.10 Simulation results: Threshold of internal materiality for resource allocation leverage. Source: Author's elaboration

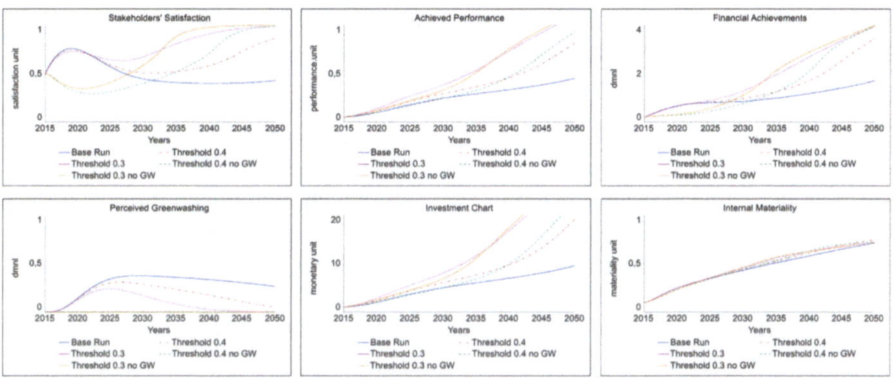

Fig. 4.11 Simulation results: Initial threshold and no greenwashing leverages. Source: Author's elaboration

behaviour. While *internal materiality* levels do not show major changes, it is the way this information is used that alters the results. Finally, scepticism also differs, as the "switch" not only favours investments but also discourages further greenwashing.

At this point in the analysis, a question arises spontaneously. What happens if greenwashing is excluded a priori?

This means combining two points of leverage—that is, greenwashing propensity and materiality threshold—instead of waiting for the "switch" to deactivate the misleading communication path. Also, this represents a very plausible scenario: Companies integrating materiality into their strategy are usually not inclined to greenwash.

To answer this question, Fig. 4.11 merges the threshold and greenwashing propensity levers, adding two additional scenarios to Fig. 4.10: the two *threshold* levels

(0.4 and 0.3) combined with the exclusion of greenwashing ("threshold 0.4 no GW", in dashed green; "threshold 0.3 no GW", in solid yellow).

In this case, the company experiences, as in every no-greenwashing scenario, an initial drop in *stakeholders' satisfaction*. However, this is compensated much faster than in Sect. 4.3.3. In the long term, the results of scenarios excluding greenwashing are better than the alternatives. Nevertheless, in the real world, it still remains difficult for a company to accept several years of underperformance compared to competitors. This explains why greenwashing practices are so widespread.

These simulations complete the analysis of the model leverage points. This complements the sensitivity results discussed in Sect. 4.2 by revealing how strategic choices can reshape system dynamics over time.

Following these findings, Section 4.4 moves beyond numerical outputs, to interpret their theoretical and managerial implications.

4.4 Discussion: Theoretical and Managerial Insights

4.4.1 Theoretical Insights

The System Dynamics analysis provides theoretical insights that enhance the understanding of sustainability reporting and its inherent tensions. These contributions build on the foundations discussed in Sect. 2.3, which highlighted the coexistence of multiple theoretical perspectives to explain the different approaches of companies to sustainability reporting (De Micco et al., 2020). Within this heterogeneous context, conflicts may emerge when companies assign different meanings to the internal and external dimensions of reporting, creating misalignments between actions and disclosures.

These theoretical perspectives have been categorised as follows, according to the weight attributed to the internal or external dimension of the sustainability reporting:

- Internal orientation: Theories such as stakeholder (Freeman, 1984), actor-network (Callon, 1984), learning-based (Albrecht et al., 2007), stewardship (Davis et al., 1997), and ecological modernisation (Spaargaren & Mol, 1992) frame sustainability reporting as a driver of organisational learning and strategic change;
- External orientation: Legitimacy (Deegan, 2002), institutional (Meyer & Rowan, 1977), and signalling (Connelly et al., 2011) theories interpret reporting as a response to external pressures and reputational concerns.

Are these views necessarily mutually exclusive, or can they be understood as complementary elements within a broader sustainability reporting paradigm?

To answer this question, it was necessary to develop a method that went beyond the predominant static approach dominating traditional sustainability reporting (Giorgino et al., 2020). System Dynamics enabled the introduction of dynamic elements into the analysis (Forrester, 1961; Meadows, 2008), starting from the definition of the concept of materiality. Materiality analysis is traditionally taught with a static, linear, and fragmented approach: The company conducts the analysis, determines the contents of the report, and publishes it, repeating this procedure, at best, every year (Garst et al., 2022).

This book applies dynamic performance management concepts to the structure of sustainability reporting, thereby overcoming these limitations (Bianchi, 2016). Materiality assumes a dynamic relevance; the significance of an issue no longer depends solely on exogenous and uncontrollable factors, but the company itself becomes a key actor in the materiality analysis. Its actions shape stakeholders' expectations, and these expectations are what determine the materiality of an issue over time.

This approach makes it possible to trace the determinants of materiality analysis and embed them within a dynamic context rich in managerial significance. Materiality is no longer a process of determining report contents (i.e., external materiality), something that remains exogenous and outside the managerial action space (Clark, 2021; Edgley, 2014). Instead, it becomes a managerial construct (i.e., internal materiality) that drives resource allocation and describes the company's overall approach to sustainability (Lai et al., 2017).

Consequently, the actors involved in the reporting process become more visible, trade-offs more manageable, and the consequences of actions clearer.

Ignoring these systemic aspects brings many problems. Overlooking the internal dimension of materiality can lead to issues of myopia, preventing the company from grasping the systemic dimension of sustainability. This myopia does not have a single dimension but can refer to several corporate aspects (see, among others, Stein, 1988). We can speak of organisational myopia, when managers fail to coordinate resources to achieve set objectives; cultural myopia, when the business model and priority system cannot adapt to a changing context; strategic myopia, when both organisational and cultural dimensions remain static, and the company fails to set new goals or organise its resources effectively; ethical myopia, when, in response to the demand for change, the company resorts to misleading practices such as greenwashing; and even cognitive-relational myopia, when the company does not know its reference environment and cannot establish effective relationships with its stakeholders.

This book explores these consequences in depth, emphasising how misalignments between the importance attributed to a sustainability issue by stakeholders and by the company can create significant tensions. Indeed, companies with low internal materiality rarely accept disclosing negative performance on issues considered critical by stakeholders (Cho et al., 2015). In such cases, companies may resort to greenwashing to satisfy stakeholders' needs (Delmas & Burbano, 2011; Jones, 2019). Hence, sustainability reporting risks losing its informational role, becoming a tool for impression management (Melloni et al., 2016). As a consequence of misleading communication, scepticism may rise, eroding trust, stakeholders' satisfaction, and subsequently affecting the company's capacity to maintain or improve its market position (Aji & Sutikno, 2015; Calabrese et al., 2015; Torelli et al., 2020). The absence of tangible results reduces the managerial relevance of sustainability, definitely inhibiting the willingness of investing for the transition.

These dynamics emerged from the "Base run" simulation (see Sect. 4.1.2), which succeeded in reproducing the system behaviour hypothesised in the problem articulation phase (see Sect. 3.2). However, the contribution of the System Dynamics method goes far beyond the simple reproduction of an expected dynamic.

First, the modelling process allows a deep understanding of the system structure. As a result, what once appeared as conflicting theories takes on a different meaning. Indeed, each theoretical lens can be used to explain a distinct phase of the reporting process—such as stakeholder engagement, resource allocation, performance management, communication practices, and the financial or impact consequences of business activities. While some lenses may be dominant in specific business contexts, it is essential to recognise that sustainability reporting is inherently complex and cannot be explained by a single, isolated perspective.

Second, understanding the system structure clarifies the systemic roots of a problem. Greenwashing practices, often considered chronic and insurmountable questions, have been endogenously defined in model terms. These are caused by misalignments between internal and external reporting purposes: When internal materiality diverges from external expectations, companies may be tempted to use greenwashing for compensating the gap between expected and actual performance. This perspective enables mapping the systemic roots of the problem and identifying effective levers to discourage misleading practices.

Third, understanding the systemic roots of a problem allows contributing to the relative academic debate. In this case, the contribution consists of offering a perspective capable of explaining the nature of an insufficiently understood phenomenon, such as the managerial role of the materiality assessment.

Indeed, the materiality debate concerning impact, financial, and double materiality approaches has dominated recent literature and represents the main point of divergence in sustainability reporting standardisation efforts. However, simulations reveal that in a dynamic context (where financial repercussions of sustainability emerge over time), the choice of materiality approach has only marginal relevance in explaining long-term system behaviour. This finding suggests that the debate, while theoretically relevant, should move from deciding which approach to adopt towards embedding materiality into strategic processes.

In this way, the System Dynamics model integrates internal mechanisms—such as materiality assessment, resource allocation, and performance management—with external perceptions, including stakeholders' expectations and financial repercussions. It therefore contributes to reconciling previously conflicting theories and perspectives, offering a boundary-spanning view where isolated problems and tensions are addressed dynamically and systemically.

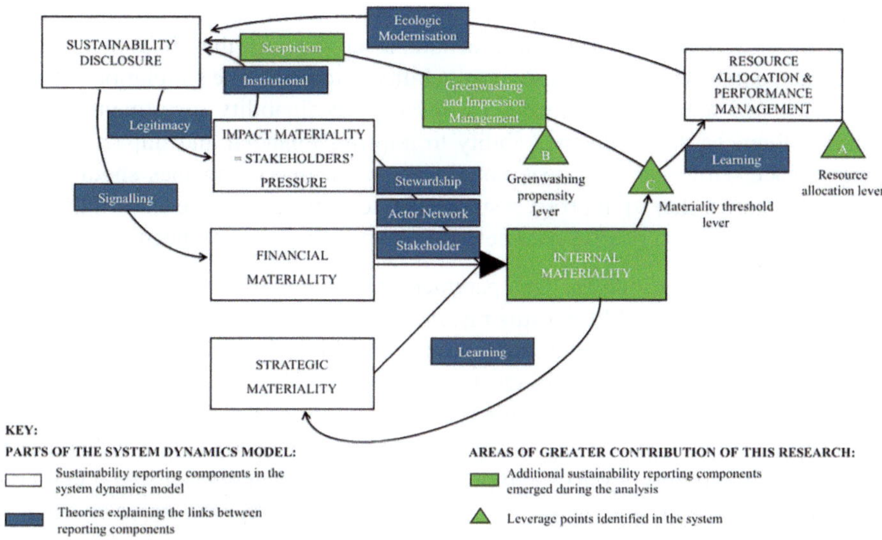

Fig. 4.12 A system perspective of sustainability reporting. Source: Author's elaboration

Figure 4.12 illustrates this systemic perspective, extending Figure 3.3 (see Sect. 3.2.3.1) by incorporating the following elements identified during the modelling process:

- Greenwashing, framed by impression management theory, as a variable influenced by low internal materiality.
- Scepticism, a consequence of greenwashing, undermining disclosure credibility.
- Leverage points identified through sensitivity analysis, marking areas for managerial intervention.

This systemic approach clarifies theoretical tensions and establishes a foundation for connecting sustainability reporting with performance management and strategic decision-making (Amigoni, 1979; Brunetti, 1979). This is particularly evident in the practical relevance of the system's leverage points, which are examined in the next subsection on the managerial implications of this book.

4.4.2 Managerial Insights

Building on the theoretical insights discussed above, the System Dynamics analysis offers practical guidance for managers aiming to leverage sustainability reporting to drive the sustainable transition.

In particular, the leverage points that emerged in the sensitivity analysis (see Sections 4.2 and 4.3) show how to increase sustainability and financial performance, while preventing undesirable dynamics, offering actionable recommendations for integrating sustainability into corporate strategy.

The first leverage point is displayed through the green triangle with the letter A in Figure 4.12. It is called resource allocation lever and describes the role played by financial resources in determining sustainability outcomes. The simulations demonstrate how the allocation of resources towards sustainability investments significantly influences the company's ability to improve financial and impact levels of performance. However, this does not simply mean that companies should invest more in sustainability to improve system behaviour; indeed, the resource allocation lever is tested for two precise parameters with specific practical meaning:

1. Initial available resources: This parameter represents the resources ready to be allocated for sustainable investments. Increasing this stock does not mean investing money immediately but being prepared to invest when conditions (i.e., the level of internal materiality) suggest doing so.
2. Normal resource allocation: This represents the speed at which resources are invested. A quick allocation strategy, especially in the early stages of the transition, is important not only to improve performance but also to signal commitment and reduce scepticism in contexts where perceived greenwashing is growing.

This leverage point suggests developing a forward-looking financial planning system. The objective of this system should be to ensure business readiness for capturing early signals of increase in the materiality and market dimension of a sustainability issue, thus allocating a sufficient quantity of resources to achieve a dominant position in the market. Once achieved, the company can count on new financial resources deriving from the positive financial repercussions of sustainability performance, ensuring continuous innovation and alignment with stakeholders' expectations to reach long-term success.

These actions can deactivate greenwashing-related loops early, enhance stakeholder trust, and create a virtuous cycle of financial returns and reinvestments. Resource allocation thus becomes a strategic lever that managers can actively control to accelerate the sustainability transition.

The second leverage point (green triangle with the letter B in Figure 4.12) refers to greenwashing propensity. This parameter captures the company's attitude towards misleading communication practices, which often emerge as a tactical response to external pressure.

Findings indicate that greenwashing produces two types of consequences: the first, appealing in the short term, satisfies stakeholder expectations quickly and enhances external legitimacy without substantial investments; the second involves the systemic risks associated with this approach. Misleading communications generate scepticism over time, eroding trust and undermining the credibility of future disclosures. Once trust is compromised, stakeholders become unresponsive to communication, even if this is grounded in genuine improvements and investments. This situation may discourage investments, further stimulating greenwashing in what has been identified as "greenwashing trap".

To prevent this situation, the model results show that excluding greenwashing can be considered a point of leverage of the system. This is not an easy choice:

Companies that avoid misleading practices suffer an initial decline in stakeholders' satisfaction, as expectations remain unmet during the early stages of the transition. However, this temporary disadvantage is offset by long-term benefits: transparency fosters trust, activates reinforcing loops of legitimacy, and accelerates the sustainability transition once performance improvements become visible. In contrast, scenarios characterised by high greenwashing propensity are dominated by persistent scepticism, which reduces financial achievements, leading to stagnation in spite of increasing market benefits.

From a managerial perspective, this leverage point highlights the importance of avoiding opportunistic communication strategies. Rather than compensating for performance gaps through greenwashing, companies should anticipate investments so as to minimise the initial drop in stakeholders' satisfaction. This approach mitigates reputational risks while strengthening stakeholder relationships, turning them into an asset for building a long-term sustainability strategy and gaining a competitive advantage.

The third leverage point (green triangle with the letter C in Figure 4.12) concerns the threshold of internal materiality for resource allocation. This parameter acts as a "switch" that determines when a company begins to invest in sustainability to improve performance instead of relying on communication strategies to compensate for external pressures. Indeed, the threshold represents the level of internal materiality that a sustainability issue must reach to acquire strategic relevance and trigger resource allocation for purchasing sustainability assets or developing green competencies.

The simulations show that lowering this threshold accelerates the transition by allowing investments to start earlier. In this way, the company moves from a reactive approach—where investments begin only when external pressure becomes evident—to a proactive one, where early signals of materiality stimulate strategic response. The functioning of this "switch" also discourages misleading communication, since investments provide tangible evidence of commitment.

However, excessively low thresholds have particular consequences: While they promote early action, they may lead to over-investment in later stages, when sustainability issues have already reached high levels of materiality. This means that the threshold requires dynamic calibration, as happens for the parameters in all performance management systems.

Finally, this leverage point must be interpreted with caution. The absence of sensitivity in the model to the type of materiality approach adopted (i.e., impact, financial, or double) might lead to the mistaken belief that materiality is not central to sustainability reporting. In reality, this leverage point signals exactly the opposite. What matters is not the approach chosen, but how the information obtained through materiality assessment is used and integrated into corporate strategy.

This is the core message of this book: Materiality analysis—regardless of the approach adopted—serves as a gateway for integrating sustainability information into strategy, whether originating from stakeholders or the market. Integrating this information into business processes is crucial to achieving a transition that is not only environmentally and socially responsible but also economically sustainable.

The presentation of the three leverage points formally concludes the analysis of the managerial implications of the simulations. However, this discussion would be truly complete only if we adopt a backward-looking orientation and ignore the potential developments and applications that this book opens up.

For this reason, the true implications of this research lie in its potential for the future: the opportunities it creates and the possible applications enabled by this generic model. These opportunities allow us to keep imagining a world where sustainability, business management, and performance measurement are considered an indivisible whole.

4.5 Conclusions: Future Applications, Opportunities, and Limitations

The model developed in this book opens several applications and opportunities that go beyond System Dynamics to enter into the research, formation, and managerial practice realms.

Understanding this potential means capturing the value added by a systemic approach to sustainability reporting, so as to overcome the fragmentation that often characterises guidelines and business practices.

For this reason, this final section shows three areas in which the research outcomes can be applied. These refer to the application of the Sustainability Reporting Model (4.5.1), the integration of the developed perspective into performance management systems (4.5.2), as well as further research avenues and limitations (4.5.3). These areas represent strategic pathways for bridging the gap between sustainability theory and business practice.

4.5.1 Potential Applications of the Sustainability Reporting Model

One of the main outcomes of this book lies in clarifying the boundaries of sustainability reporting, which often prove difficult to be defined and fully understood. In particular, the model helps distinguish the internal and external dimensions of reporting—represented by internal materiality, conceptualised as a strategic lever, and external materiality, identified through stakeholders' pressure—and shows how these dimensions are interconnected. Understanding the system structure becomes especially important in dynamic contexts, where business choices influence and are influenced by stakeholders' expectations.

This conceptual clarification opens new opportunities for scientific and educational dissemination. The model can serve as a teaching tool in academic seminars, practitioner workshops, and training courses, both in-person and online. Simulations allow participants to explore, in interactive and participatory ways, the consequences of alternative managerial choices. Moreover, it is possible to shed light on leverage points of the system that are often overlooked in mainstream sustainability discourse, such as internal materiality or greenwashing propensity. As shown in this book, these elements hold greater potential to generate systemic consequences over the long term than the chosen materiality approach. Yet, dissemination initiatives and debates increasingly focus on impact, financial, and double materiality approaches only, without addressing how to strategically integrate these perspectives into corporate decision-making.

As said, these activities can take different forms:

- Academic seminars provide opportunities to interact with other scholars, identify new areas for model integration, and explore research avenues that expand the impact of this work.
- Practitioner workshops can foster change within business contexts, challenging existing routines and demonstrating how the transition can also be economically sustainable.
- Online and offline training courses encourage active participation, whether through interactive platforms that allow users to simulate alternative business decisions or through in-person discussions of specific scenarios, aimed at identifying leverage points and testing system responses to change.

Beyond dissemination, the model can act as a bridge between research and business practice. During workshops or courses, it can be tailored to test strategic hypotheses and evaluate the impact of different investment policies across sustainability dimensions. In this way, the model becomes a virtual laboratory for scenario analysis, enabling companies to assess potential risks associated with more or less aggressive investment strategies before translating them into operational decisions.

The model also offers a powerful tool for raising awareness about greenwashing risks. Simulations can reproduce scenarios where companies attempt to reduce external scepticism and regain legitimacy, highlighting the boomerang effect of misleading practices. For example, step-by-step simulations may initially stimulate greenwashing by providing a competitive advantage at low cost; however, this advantage is soon eroded by rising scepticism, with negative consequences for long-term financial performance and corporate legitimacy. Using the model in educational contexts makes this dynamic tangible, especially when combined with visual or interactive experiences that illustrate real-world cases of exaggerated communication.

Moving to the managerial realm, the operational potential of the model lies in its customisability. Starting from its generic structure, it can be adapted to a specific business context by mapping stakeholders and assessing their relevance in terms of both impact and financial dimensions. This process involves evaluating the influence of each stakeholder group across sustainability dimensions, thereby expanding the model into a more detailed representation tailored to the organisation's complexity and uniqueness. Such adaptation transforms the model into a practical tool for strategic planning, facilitating the integration of materiality analysis with resource allocation and investment prioritisation.

However, it would be reductive to limit these opportunities to the restricted community of System Dynamics users. This book was conceived with a boundary-spanning objective: to bridge the gap between theory and practice. For this reason, the following subsection illustrates how the research outcomes can be translated into performance management procedures without necessarily relying on dynamic simulation.

4.5.2 *Lessons Learned and Integration into Performance Management and Control Systems*

The dynamics emerging from the simulations of the Sustainability Reporting Model offer valuable insights for managerial practice, particularly regarding the mechanisms that explain relationships between companies and stakeholders and the roots and consequences of greenwashing. However, applying the model "as is" to a real case could lead to unintended distortions.

For example, adapting the model would require estimating the benefits of greenwashing or the time lag before scepticism arises. This could generate undesirable effects, such as searching for "optimal" levels of greenwashing or identifying the best moment to stop opportunistic practices before distrust becomes irreversible. Such uses would contradict the model's purpose: raising awareness of the systemic implications of sustainability reporting and discouraging opportunistic behaviour.

Therefore, practical applications of this research do not necessarily involve replicating the System Dynamics simulation. What matters is that users understand the components of the sustainability reporting system and know how to apply these conceptual foundations using familiar tools and methods.

In particular, the model structure highlights areas that should be integrated into corporate performance management systems. This means embedding sustainability within existing control frameworks, adding measurement dimensions, and overcoming the traditional separation between financial and non-financial reporting.

To achieve this integration, companies can develop a structured process articulated in the following steps:

1. Stakeholder identification: Map the most influential actors who impact (or are impacted by) business activities. Assess each group's relevance in terms of pressure and legitimacy.
2. Impact materiality analysis: Evaluate the most significant sustainability issues for each stakeholder group. For each issue, define a desired performance level based on stakeholders' expectations, weighted by their relevance. This constitutes the sustainability benchmark to which to aspire.
3. Financial materiality analysis: Estimate the financial repercussions of achieving sustainability performance, summing positive and negative impacts associated

with different performance levels. This clarifies the strategic importance of each issue.
4. Materiality index determination: For each topic, combine impact and financial dimensions, assigning weights aligned with the company's values. Integrate the resulting materiality indicator into budgeting and resource allocation processes.
5. Diffusion of materiality across the organisation: Extend materiality assessment beyond sustainability issues to all business activities, creating a unified measurement system. This allows categorisation of activities by internal materiality.
6. Full integration into decision-making: Prepare a list of actions and improvements for each category, simulating their effects on performance. Conduct cost-benefit analyses for each investment and compare results with stakeholders' expectations. Integrate the outcomes into budgeting and planning systems.
7. Strategic review: Evaluate the proposed investment plan against business priorities and resources. Translate materiality analysis into operational planning and reassess strategy in light of emerging sustainability objectives.

This process should be adjusted annually, comparing results with targets and updating the materiality analysis. The outcome is a sustainability performance management framework fully integrated with the business model, materiality analysis, and reporting process.

Although this framework lacks some benefits of System Dynamics—such as simulating stakeholder pressure or predicting issue evolution—it still represents a critical step forward if compared with current business practices. Although in concrete terms, annual adjustment ensures a dynamic approach to sustainability management, even without simulation.

In conclusion, while System Dynamics offers advantages in feedback and scenario analysis, its conceptual foundations can be translated into operational tools for performance management and control. This integration is essential for embedding sustainability into decision-making, transforming reporting from a compliance exercise into a strategic lever for long-term value creation.

4.5.3 Limitations of the Study and Future Research Opportunities

After having discussed the implications of the study for both theory and practice, it is now necessary to acknowledge some limitations linked to the research method and the model structure. From a System Dynamics perspective, the Sustainability Reporting Model presents constraints that affect its applicability to real cases and its generalisability, such as:

- Reliance on intangible soft variables: Many causal relationships are based on aspects that are difficult to quantify, such as stakeholders' expectations or levels of materiality and satisfaction. This choice, while necessary to represent all dimensions of sustainability reporting, introduces a high degree of subjectivity, which can limit real-world applicability.

- Aggregated representation of stakeholders: The model simplifies company–stakeholder relationships by assuming that all stakeholders can be aggregated into a single homogeneous class. This ensures a generic, context-independent structure but requires significant adaptation for specific environments.
- Absence of market and issue specifics: The generic perspective overlooks details that vary significantly across markets and sustainability issues. Estimating these for real cases would require extensive data collection.
- Simplified relationships between materiality and performance: Connections between internal and external materiality, as well as financial repercussions, are modelled through easy mathematical functions that may not fully capture the non-linear nature of these phenomena.
- Exclusion of relevant variables from model boundaries: The model does not consider factors such as competitor behaviour, the impact of other companies' greenwashing on general scepticism, assurance requirements to enhance reporting reliability, or the negative consequences of insufficient sustainability performance. These elements, driven by exogenous forces, are difficult to define in System Dynamics terms but have significant real-world impacts.

Acknowledging these limitations opens up new research avenues. First, there is space for improving the model structure by estimating more complex equations and expanding system boundaries to enable more precise analyses. Furthermore, integrating empirical data with multi-modelling techniques (e.g., combining System Dynamics with Agent-Based Modelling) could better capture actor diversity (such as different stakeholder groups) and their dynamic interactions.

Future research could focus on:

- Developing sector-specific versions of the model calibrated for particular industries.
- Exploring the role of assurance practices and public policies in shaping systemic outcomes and mitigating greenwashing risks.
- Analysing interactions between stakeholders and companies, including competitive dynamics.
- Testing the integration of the model into performance management systems to assess its operational validity.

In summary, the proposed model represents an initial step towards understanding the dynamics generated by sustainability reporting activities. Its evolution will require further methodological and practical efforts to create a ready-to-apply tool capable not only of stimulating managerial reasoning but also of providing specific support for the sustainability transition.

References

Aji, M. H., & Sutikno, B. (2015). The extended consequence of greenwashing: Perceived consumer skepticism. *International Journal of Business and Information, 10*(4).

Albrecht, P., Burandt, S., & Schaltegger, S. (2007). Do sustainability projects stimulate organizational learning in universities? *International Journal of Sustainability in Higher Education, 8*(4), 403–415. https://doi.org/10.1108/14676370710823573

Amigoni, F. (1979). *I sistemi di controllo direzionale: criteri di progettazione e di impiego*. Giuffrè.

Bianchi, C. (2016). *Dynamic performance management* (Vol. 1). Springer International Publishing. https://doi.org/10.1007/978-3-319-31845-5

Brunetti, G. (1979). *Il controllo di gestione in condizioni ambientali perturbate*. FrancoAngeli.

Calabrese, A., Costa, R., & Rosati, F. (2015). A feedback-based model for CSR assessment and materiality analysis. *Accounting Forum, 39*(4), 312–327. https://doi.org/10.1016/j.accfor.2015.06.002

Callon, M. (1984). Some elements of a sociology of translation: Domestication of the scallops and the fishermen of St Brieuc Bay. *The Sociological Review, 32*(1_suppl), 196–233. https://doi.org/10.1111/j.1467-954X.1984.tb00113.x

Cho, C. H., Laine, M., Roberts, R. W., & Rodrigue, M. (2015). Organized hypocrisy, organizational façades, and sustainability reporting. *Accounting, Organizations and Society, 40*, 78–94. https://doi.org/10.1016/j.aos.2014.12.003

Clark, C. E. (2021). How do standard setters define materiality and why does it matter? *Business Ethics, the Environment & Responsibility, 30*(3), 378–391. https://doi.org/10.1111/beer.12351

Connelly, B. L., Certo, S. T., Ireland, R. D., & Reutzel, C. R. (2011). Signaling theory: A review and assessment. *Journal of Management, 37*(1), 39–67. https://doi.org/10.1177/0149206310388419

Davis, J. H., Schoorman, F. D., & Donaldson, L. (1997). Toward a stewardship theory of management. *Academy of Management Review, 22*(1), 20–47. https://doi.org/10.5465/amr.1997.9707180258

De Micco, P., Rinaldi, L., Vitale, G., Cupertino, S., & Maraghini, M. P. (2020). The challenges of sustainability reporting and their management: The case of Estra. *Meditari Accountancy Research, 29*(3), 430–448. https://doi.org/10.1108/MEDAR-09-2019-0555

Deegan, C. (2002). Introduction: The legitimising effect of social and environmental disclosures—A theoretical foundation. *Accounting, Auditing & Accountability Journal, 15*(3), 282–311. https://doi.org/10.1108/09513570210435852

Delmas, M. A., & Burbano, V. C. (2011). The drivers of greenwashing. *California Management Review, 54*(11).

Edgley, C. (2014). A genealogy of accounting materiality. *Critical Perspectives on Accounting, 25*(3), 255–271. https://doi.org/10.1016/j.cpa.2013.06.001

Forrester, J. W. (1961). *Industrial dynamics*. Massachusetts Institute of Technology (MIT).

Freeman, R. E. (1984). *Strategic management. A stakeholder approach*. Pitman.

Garst, J., Maas, K., & Suijs, J. (2022). Materiality assessment is an art, not a science: Selecting ESG topics for sustainability reports. *California Management Review, 65*(1), 64–90. https://doi.org/10.1177/00081256221120692

Giorgino, M. C., Barnabè, F., & Kunc, M. (2020). Integrating qualitative system dynamics with accounting practices: The case of integrated reporting and resource mapping. *Systems Research and Behavioral Science, 37*(1), 97–118. https://doi.org/10.1002/sres.2602

Jones, E. (2019). Rethinking greenwashing: Corporate discourse, unethical practice, and the unmet potential of ethical consumerism. *Sociological Perspectives, 62*(5), 728–754. https://doi.org/10.1177/0731121419849095

Lai, A., Melloni, G., & Stacchezzini, R. (2017). What does materiality mean to integrated reporting preparers? An empirical exploration. *Meditari Accountancy Research, 25*(4), 533–552. https://doi.org/10.1108/MEDAR-02-2017-0113

Lewicki, R. J., & Bunker, B. B. (1996). Developing and maintaining trust in work relationships. In R. M. Kramer & T. R. Tyler (Eds.), *Trust in organizations: Frontiers of theory and research* (pp. 114–139). Sage Publications.

Meadows, D. H. (2008). *Thinking in systems* (Sustainability Institute, Ed.).

Melloni, G., Stacchezzini, R., & Lai, A. (2016). The tone of business model disclosure: An impression management analysis of the integrated reports. *Journal of Management and Governance, 20*(2), 295–320. https://doi.org/10.1007/s10997-015-9319-z

Meyer, J. W., & Rowan, B. (1977). Institutionalized organizations: Formal structure as myth and ceremony. *American Journal of Sociology, 83*(2), 340–363. https://doi.org/10.1086/226550

Slovic, P. (1993). Perceived risk, trust, and democracy. *Risk Analysis, 13*(6), 675–682.

Spaargaren, G., & Mol, A. P. J. (1992). Sociology, environment, and modernity: Ecological modernization as a theory of social change. *Society & Natural Resources, 5*(4), 323–344. https://doi.org/10.1080/08941929209380797

Stein, J. C. (1988). Takeover threats and managerial myopia. *Journal of Political Economy, 96*(1), 61–80.

Torelli, R., Balluchi, F., & Lazzini, A. (2020). Greenwashing and environmental communication: Effects on stakeholders' perceptions. *Business Strategy and the Environment, 29*(2), 407–421. https://doi.org/10.1002/bse.2373

Appendix "A": Model Testing

This appendix presents the various tests conducted to ensure the validation and robustness of the model, evaluating its consistency, accuracy, and reliability in reflecting the underlying dynamics of sustainability reporting and decision-making processes.

The literature analysis performed in Chapter 2 provided scientific roots for the model, ensuring a first type of structure confirmation test (Sterman, 2000, p. 863).

The parameters and variables (Sterman, 2000, p. 867) contained in the model have real-world counterparts, as confirmed by the explanatory names used. The relationships between different concepts are usually assumed to be linearly increasing or decreasing according to the relative polarity. The three cases of non-linearity ("effect of investments on performance", "effect of financial on internal materiality", and "perceived greenwashing multiplier on communicated performance") are represented through graphical functions and specifically described and supported by the literature (Aji & Sutikno, 2015; Frankel, 1962; IFRS, 2023).

Dimensional consistency (Sterman, 2000, p. 866) is ensured by attributing units to most of the variables. Generic "satisfaction", "materiality", and "performance" units have been adopted for the soft stocks. The relative nature of financial achievement levels suggested the use of dimensionless (dmnl) units; differently, "monetary unit" was chosen for the asset level, which could have both a relative (% of total assets) and a precise (amount of invested money) meaning. Greenwashing represents the quantity of performance shortfall compensated through misleading communication; coherently, it is measured in "performance units". No unit errors are signalled by the software (Stella Architect 3.8).

Direct extreme-condition tests (Sterman, 2000, p. 869) were performed thanks to the division of the model into five sectors and ensured the model's solidity. MIN, MAX, and SAFEDIV functions were added to support solidity and maintain the original sense of the variables. Detailed explanations were provided for each one of these functions. The model exhibited plausible responses when exposed to extreme values. For example, Fig. A.1 shows the results achieved with the following settings:

average global materiality = 0, *initial stakeholders' satisfaction* = 1, *initial sustainability dedicated asset and competencies* = 1.

R. Santoni, *The Dynamics of Materiality*, System Dynamics for Performance Management & Governance 8, https://doi.org/10.1007/978-3-032-19280-6

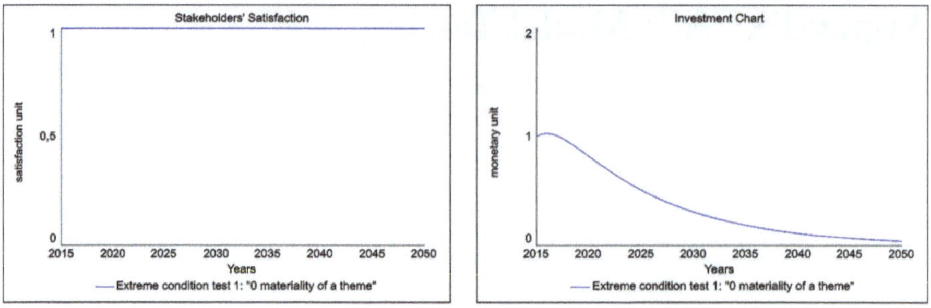

Fig. A.1 Extreme condition test 1: No materiality of an issue. Source: Author's elaboration

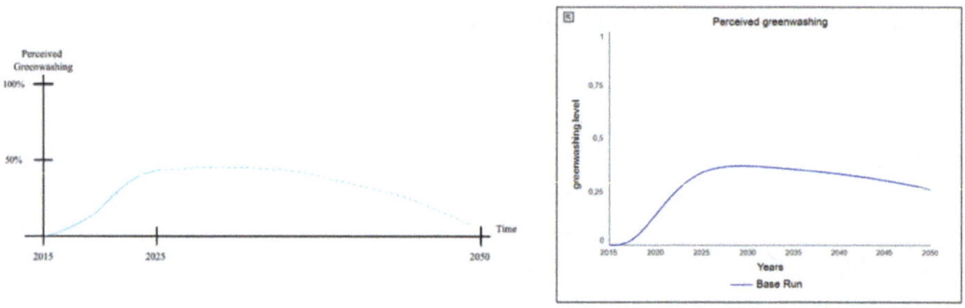

Fig. A.2 Reproduction of reference mode of behaviour: The variable "perceived greenwashing". Source: Author's elaboration

This scenario simulated a situation in which no interest exists in a certain sustainability theme (e.g., the explosion of the Sun); the absence of interest means a complete satisfaction of stakeholders, independently from the company's behaviour, and zero performance expectations; interestingly, the initial conditions simulated a certain level of internal materiality (e.g., a "creative" manager) supported by some active investments in the assets and competencies dedicated to this unusual sustainability challenge. Coherently, after the materiality assessment, the model portrays the interruption of investments, with the stock gradually decreasing according to its depletion time.

As concerns integration error tests (Sterman, 2000, p. 872) the simulation was run with the following settings: Euler method, timestep = 0.125 (1/8). No changes in behaviour emerged by adopting a smaller DT or moving to a different integration method.

Reference mode reproduction (Sterman, 2000, p. 874). As shown in Fig. A.2, the results of the "Base run" (on the right) are similar – even though they show a greater scepticism resistance – to the expected behaviour of the perceived greenwashing variable (on the left), previously depicted in Figure 3.2 (Sect. 3.2.2).

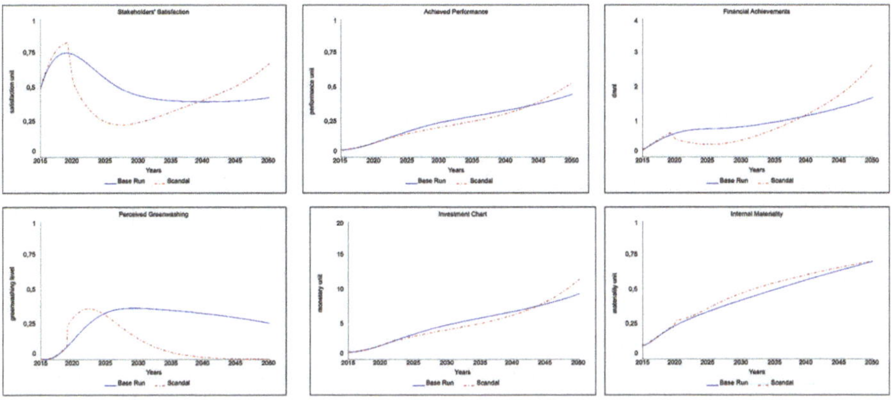

Fig. A.3 Behaviour reproduction tests: Simulating a green scandal. Source: Author's elaboration

An additional behaviour reproduction test was performed by simulating a scandal involving a greenwashing firm. The chosen scenario was the following: in 2019, a scandal involved a firm that was greenwashing. All of a sudden, all the newspapers reported data falsification on that firm's environmental emissions. The spread of awareness made people angry and dissatisfied; the firm needed to react immediately to avoid collapse and regain people's trust.

To simulate that event, the greenwashing perception delay was set to zero at the moment of the scandal. In model terms, this meant decreasing for 1 year (2019–2020), through "step" and "pulse" functions (Sterman, 2000, pp. 413, 426), the time delays driving *perceived greenwashing*, *change in stakeholders' satisfaction*, *change in stakeholders' materiality*, and *change in internal materiality*. DT was put to 1/128. This favoured a quick change in stakeholders' perception of the firm's behaviour and ethics, determining a drop in satisfaction. Right after the scandal, the firm strategy changed; the greenwashing strategy was abandoned (*greenwashing propensity* initially set to 1, then moved to 0 with a "step" function), and *internal materiality* became mainly determined by impact materiality instead of by the level of financial achievements (*weight of financial materiality* changed from 0.5 to 0.25 from 2019). The results in Fig. A.3 show how a loss in credibility (i.e., an increase in *perceived greenwashing* and a drop in *stakeholders' satisfaction*) requires a long time to be counteracted, even when allocating more available resources and having higher *internal materiality* levels. From a financial achievement perspective, this scandal has determined years of underperformance, before finally reaching an improved scenario in the last 10 years of simulation.

Although not originally designed for this purpose, the model produced plausible results under the simulated stress conditions.

The last category of tests is the sensitivity analysis (Sterman, 2000, p. 883) of the model's parameters. Its main results, concerning variables showing policy sensitivity (see Sect. 4.2) acted as a bridge between the testing and the findings phases.

The part of sensitivity analysis portrayed here concerns variables with "behavioural" sensitivity, i.e., areas of model uncertainty that do not constitute space for managerial intervention. This analysis is useful in explaining the choices behind certain parameter values or table function shapes, constituting an essential part of the model validation.

The first case of behavioural sensitivity portrayed in this appendix concerns the value of *initial satisfaction*, which depends on the initial business reputation and causes different system behaviours, depicted in Fig. A.4.

Between 0 and 0.75, the scenario changes only numerically; differently, when the initial satisfaction is at its maximum (i.e., 1) the loss of reputation due to scepticism is balanced by resilience and customer loyalty.

Other cases of behavioural sensitivity are represented by scenarios in which more assets are required to reach the same level of performance, as shown in Fig. A.5 (i.e., parameters *normal dedicated investments*, and *normal performance level*). The same asset level would lead to worse performance, decreasing satisfaction and stimulating greenwashing. When the *normal dedicated investments* is lower (e.g., 3), the same quantity of assets is more performant and leads to higher, maintained satisfaction levels.

The last sensitivity analysis of the sector concerns the variable *perceived greenwashing multiplier on satisfaction*. The effects on *stakeholders' satisfaction* created by different shapes of the graphical function are shown in Fig. A.6. The results confirm the expectations: if the effects of scepticism are slower or smaller, the levels of satisfaction are higher. In particular, the base settings (Fig. A.5) present a situation in which, when the stakeholders identify the 10% of total communication as misleading (i.e., *perceived greenwashing*), they will not trust around 15% of the total information. This means that not only the communication perceived as misleading is distrusted, but also an additional amount, for scepticism and suspicion reasons (Fein & Hilton, 1994; Kramer, 1999). Anyhow, the model showed only numerical sensitivity (i.e., higher satisfaction but similar trends) when excluding suspicion effects and representing the distrust through a linear function (meaning: the stakeholders do not trust only the perceived greenwashing information, without

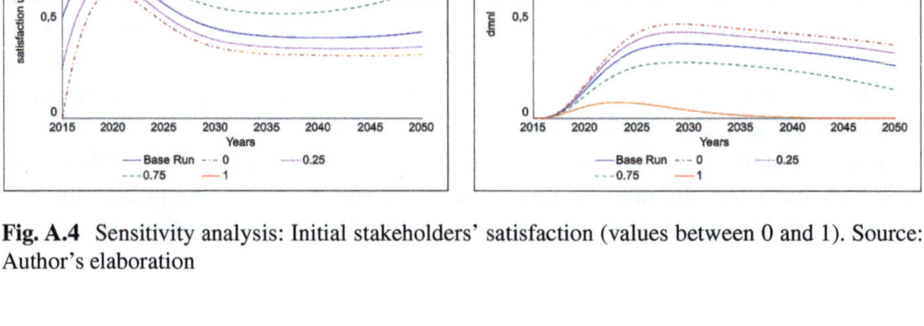

Fig. A.4 Sensitivity analysis: Initial stakeholders' satisfaction (values between 0 and 1). Source: Author's elaboration

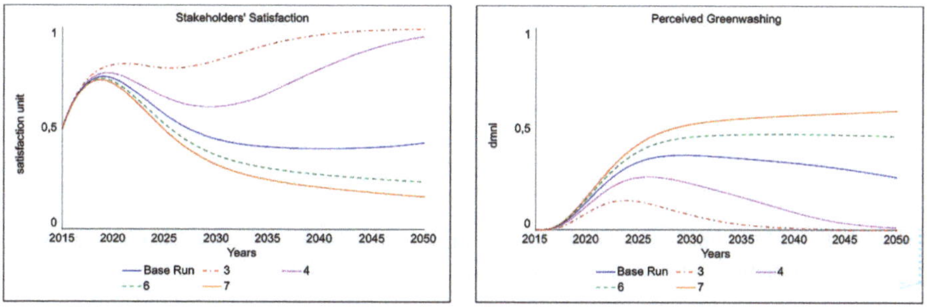

Fig. A.5 Sensitivity analysis: Normal dedicated investments (values between 3 and 7). Source: Author's elaboration

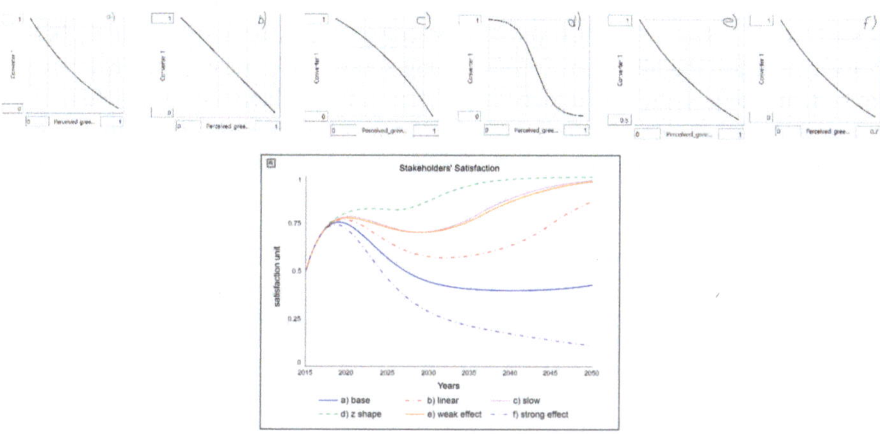

Fig. A.6 Sensitivity analysis: Perceived greenwashing multiplier on communication. Source: Author's elaboration

scepticism on the other communications), (Fig. A.6). Differently, behavioural sensitivity emerges when assuming alternative conditions (Fig. A.6). These formulations assume that stakeholders continue to believe the business communication even when they perceive greenwashing (c), or when the greenwashing is under a certain threshold (d, z-shape function), or that they still trust 50% of the communication even when the 100% is perceived as greenwashing (e, higher minimum value of the function). The last experiment (Fig. A.6) tests the situation in which 70% of perceived greenwashing is enough to lose 100% of trust in business communication characterised by a numerical sensitivity of lower satisfaction levels.

Moving to the stakeholders' materiality sector, the exogenous *average global materiality* level, modelled as a time-dependent table function, acts as the external driver of stakeholders' sense of concern, with the company that needs to react to increasing levels of apprehension related to sustainability issues. What if global

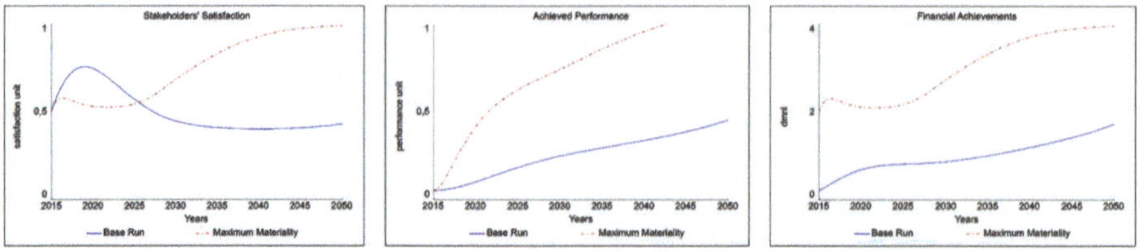

Fig. A.7 Sensitivity analysis: Average global materiality—base scenario (linear increase) and maximum materiality (constant value of 1). Source: Author's elaboration

materiality stays at its maximum since the beginning, when the business is not ready to perform?

Figure A.7 compares the base scenario (of constantly growing global materiality) with an alternative scenario where global materiality begins and remains at its maximum level (1). Here, the firm quickly perceives the issue's significance, begins investing earlier and abandons greenwashing sooner. As a result, the behaviour of the system changes: the firm understands that a shift in behaviour is required as soon as possible, *stakeholders' satisfaction* reaches its peak rapidly, driven by levels of *achieved performance* that exceed the upper target of 1. This, in turn, drives to the maximisation of financial achievements, in a market fully oriented toward sustainability.

This set of tests demonstrated the model's response to alternative settings and initial conditions. The plausibility of the generated scenarios confirms the robustness of the model and offers valuable insights that may inspire further exploration.

Appendix "B": Model Documentation

This appendix contains the complete model documentation, including variable names, equations, units, and software settings, to ensure transparency and facilitate research reproducibility. The model was developed and implemented using Stella Architect, version 3.8 (Tables B.1 and B.2).

Table B.1 Model documentation

Variable name	Equation	Units	Documentation
"1—Internal materiality sector":			
Change in internal materiality	(reference internal materiality * cumulative effect on internal materiality—internal materiality) / time to adjust internal materiality	Materiality unit/years	The flow change in internal materiality is responsible for the adjustment of the stock over time. It represents a first-order information delay structure where the actual information (i.e., reference internal materiality * cumulative effect on internal materiality) acts as a goal to which the previous level of the stock (i.e., internal materiality) tends to align. The delay to close the gap is determined by the time to adjust internal materiality.

(continued)

R. Santoni, *The Dynamics of Materiality*, System Dynamics for Performance Management & Governance 8, https://doi.org/10.1007/978-3-032-19280-6

Table B.1 (continued)

Variable name	Equation	Units	Documentation
Cumulative effect on internal materiality	Effect of financial on internal materiality * weight of financial materiality + effect of impact on internal materiality * (1—weight of financial materiality)	dmnl	The factor cumulative effect on internal materiality summarises, in a value between 0 and 2, the effects of the impact and financial components on internal materiality. In this way, the product of the multiplication with the reference internal materiality (0.5) can only assume values between 0 and 1, respecting the minimum and maximum requirements of the internal materiality stock.
Effect of financial on internal materiality	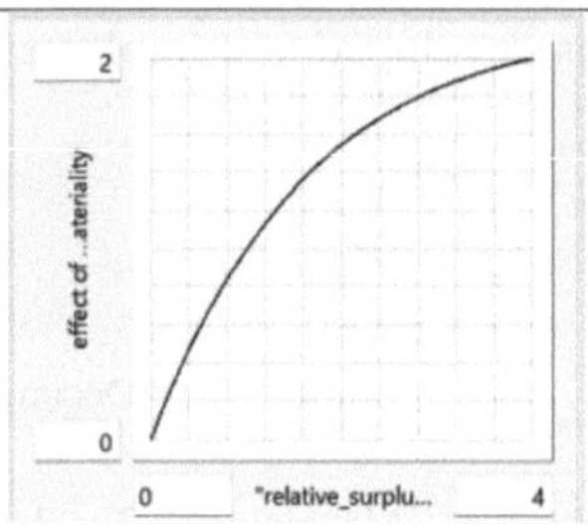	dmnl	This table function portrays the effect (Y-axis ranging between 0 and 2) that the financial achievements (i.e., the measure of financial materiality which can range between 0 and 4 on the X-axis) has on internal materiality. The function has a decreasing angular coefficient, due to the higher responsiveness of companies to the first signals of market opportunities; for this, the function passes through the points (0;0), (1;1), and (4; 2).
Effect of impact on internal materiality	Relative stakeholders' materiality * sensitivity of internal materiality to impact materiality	dmnl	The impact dimension of materiality depends on the level of materiality for stakeholders, which is modelled similarly to the internal materiality stock. For this reason, the effect of impact on internal materiality is based on a linear relationship.

(continued)

Table B.1 (continued)

Variable name	Equation	Units	Documentation
Initial internal materiality	0,1	Materiality unit	The initial value of the internal materiality stock is set to 0.1 materiality units: It is assumed that at the beginning of the simulation, the sustainability issue plays a marginal role in business decision-making, with a 10% materiality score.
Internal materiality(t)	Internal materiality(t—dt) + (change in internal materiality) * dt	Materiality unit	The materiality of an issue for an organisation (i.e., internal materiality) is an accumulation of information over time and represents the level of priority attributed to a determined sustainability issue (Ferrero-Ferrero et al., 2018; OIBR, 2022). It is represented as a soft variable stock that can assume values between 0 and 1.
Reference internal materiality	0,5	Materiality unit	The reference internal materiality is set to 0.5 materiality units (i.e., 50%) as the medium value of materiality on which to calculate the cumulative effect on internal materiality.
Sensitivity of internal materiality to impact materiality	1	dmnl	The sensitivity of internal materiality to impact materiality is the change in the effect according to a unitary change in the relative stakeholders' materiality. Due to the common nature of the two stocks, this sensitivity variable has a value of 1.

(continued)

Table B.1 (continued)

Variable name	Equation	Units	Documentation
Time to adjust internal materiality	2	Year	The time to adjust internal materiality is set to 2 years: guidelines require to annually perform the materiality assessment; however virtuous companies declared to perform it on average every 1.4 years (OIBR, 2022, p. 33). A longer time must be considered when the materiality assessment concerns the development or the revision of the sustainability strategy (Garst et al., 2022, p. 81), like in the context of this research.
Weight of financial materiality	0,5	dmnl	The variable weight of financial materiality is used to portray the financial, impact, or double approach to materiality of the analysed company. A double materiality orientation is simulated by attributing a value of 0.5; to simulate impact approaches a value of 0.25 is suggested, while for financial ones use 0.75.
“2—Resource allocation sector”:			
Asset depletion	Sustainability dedicated assets and competencies / asset depletion time	Monetary unit/year	This rate represents the value lost every year by the stock of assets and competencies due to depletion and obsolescence.
Asset depletion time	10	Years	An average life of 10 years keeps count of the obsolescence rate of the sustainability investments.

(continued)

Table B.1 (continued)

Variable name	Equation	Units	Documentation
Assets and competencies per financial achievement unit	1	Monetary unit/year	This variable represents the quantity of sustainable assets (measured in monetary terms) that can be bought by investing an amount of financial achievement units equal to 1.
Available resources allocation	(available resources * effect of internal materiality on available resource allocation*normal percentage of resources) / investment time	Monetary unit/year	This flow represents the available resources (not derived from financial achievements) that are allocated every year to increase the quantity of sustainability assets and competencies.
Available resources(t)	Available resources(t—dt) + (− available resources allocation) * dt	Monetary unit	This stock represents the resources that a firm can allocate to a certain sustainability issue. It can derive from internal savings or external sources, like bank loans. Its value is measured as a percentage of the total assets owned by a firm but could also assume a monetary meaning (e.g., ten million dollars). For this reason, it is measured in monetary terms.Its initial value of 10 is the result of calibration. It is assumed that a firm has the potential to double the normal asset level.
Effect of internal materiality on available resource allocation	1 + resource allocation sensitivity to internal materiality * (relative internal materiality—1)	dmnl	The equation is structured in the form of Y = 1 + A (X − 1), where Y is the effect, A is the constant parameter (i.e., the sensitivity) and X is the relative materiality.

(continued)

Table B.1 (continued)

Variable name	Equation	Units	Documentation
Effect of internal materiality on greenwashing	1+ sensitivity of greenwashing to internal materiality * MIN(1; (relative internal materiality-1))	dmnl	The structure of this equation is Y = 1 + A (X-1). The A parameter, the sensitivity, is negative and equal to −1. The MIN function is used to prevent anomalies (i.e., greenwashing incentives below 0) during the sensitivity tests in which relative materiality goes above 2.
Effect of materiality on reinvestments	MIN(1; relative internal materiality * sensitivity of reinvestments to materiality)	dmnl	If the firm's materiality increases, the firm will consider staying in the market as essential. It will reinvest a higher percentage of financial achievements in assets to maintain or improve its performance.The minimum function (i.e., MIN) tells the software to select the minimum value among two or more possible equation outputs. In this case, it prevents the possibility of reinvesting more financial achievements than achieved when simulations are made with a threshold level below 0.5.
Greenwashing incentive	MIN(1; (greenwashing propensity * effect of internal materiality on greenwashing))	dmnl	The variable greenwashing incentive represents the percentage of performance shortfall—as compared with stakeholders' expectations—that the company compensates through some form of greenwashing (Delmas & Burbano, 2011; Jones, 2019) in normal conditions.

(continued)

Table B.1 (continued)

Variable name	Equation	Units	Documentation
Greenwashing propensity	0,5	dmnl	The greenwashing propensity, set to 0.5, indicates that the 50% of performance shortfall is compensated when the materiality equals the threshold.
Initial available resources	10	Monetary unit	This initial value of the stock of available resources is set to 10 as a result of calibration. It is assumed that a firm has the potential to double the normal assets level.
Initial sustainability asset and competencies	1	Monetary unit	In the beginning, the firm's commitment to sustainability is low and is reflected by the initial low level of investments equal to 1, representing the 4% of the endowment necessary to complete the transition.
Investment time	3	Year	Three years is the normal duration of a business or investment plan.
Normal percentage of resources	0,25	dmnl	This parameter indicates a percentage of the total resources in the stock that is invested in topic-specific assets and competencies during the investment time.A value of 0.25 means that a firm desires to invest (during in the investment time) 25% of the available resources when materiality = threshold.
Reinvested financial achievements	Financial achievements (or financial materiality)*effect of materiality on reinvestments	dmnl	This variable represents the percentage of financial achievements reinvested according to the comparison between internal materiality and the threshold.

(continued)

Table B.1 (continued)

Variable name	Equation	Units	Documentation
Reinvestments	Reinvested financial achievements*assets and competencies per financial achievement unit	Monetary unit/year	Yearly resources that the firm reinvests to increase assets and competencies.
Relative internal materiality	Internal materiality/threshold of internal materiality for resource allocation	dmnl	The relative internal materiality is obtained by comparing the level of the internal materiality stock to a threshold of internal materiality for resource allocation.
Resource allocation sensitivity to internal materiality	1	dmnl	This parameter shows the linear relationship between materiality change and extra resource allocation.
Sensitivity of greenwashing to internal materiality	-1	dmnl	This parameter represents the negative linear relationship between materiality and greenwashing incentive. The MIN function is used to prevent the greenwashing incentive from going above 1 when values of the threshold above 0.5 are used in the simulation.
Sensitivity of reinvestments to materiality	0,5	dmnl	The variable sensitivity of reinvestments to materiality has a value of 0.5, making the effect (i.e., the multiplier) ranging between 0 and 1
Sustainability dedicated assets and competencies(t)	Sustainability dedicated assets and competencies(t—dt) + (reinvestments + available resources allocation—asset depletion) * dt	Monetary unit	This stock is valuated in monetary units, representing the set of assets and competencies dedicated to the sustainability issue under investigation. Like all investments, the stock is subject to a depletion process.

(continued)

Table B.1 (continued)

Variable name	Equation	Units	Documentation
Threshold of internal materiality for resource allocation	0,3	Materiality unit	The threshold of internal materiality for resource allocation is the level of materiality that a company perceives as necessary to attribute strategic meaning to the sustainability issue, so as to start allocating resources to improve the related performance. It is initially set to the neutral value of 0.5, equal to the reference internal materiality.
"3—Performance communication sector":			
Achieved performance	Normal performance level * effect of investments on performance	Performance unit	Performance level deriving from the level of sustainability dedicated assets and competencies. It is measured in performance units. When a value of 1 is reached, the sustainability transition is completed. Values above 1 mean results over the requirements.
Change in satisfaction	(indicated satisfaction—stakeholders' satisfaction) / time to adjust satisfaction	Satisfaction unit/years	This biflow changes the value of the stock of satisfaction. It is formulated as an information delay between the indicated level of satisfaction and the actual value of the stock. Time delay is required to adjust perception.
Communicated performance	(achieved performance + greenwashing performance communication)	Performance unit	The sum of achieved performance and greenwashing performance communication constitutes the communicated performance, i.e., the contents of the report. From this moment on, the information becomes available also to stakeholders.

(continued)

Table B.1 (continued)

Variable name	Equation	Units	Documentation
Communication adequacy	MIN(1; SAFEDIV(communicated performance; stakeholders' expected performance; 1))	dmnl	This variable measures the ratio between communicated performance and stakeholders' expectations. It affects the stakeholder's perceived adequacy of the communication.The MIN function prevents adequacy from going above 1 when a firm communicates better performance than required.When the expected performance is 0, the adequacy is supposed to be 1.
Communication effect on satisfaction	Satisfaction sensitivity to communication adequacy * communication adequacy*perceived greenwashing multiplier on communication	dmnl	The satisfaction level is positively impacted by the net communication adequacy, function of the total communication and its reliability (effect of perceived greenwashing).
Effect of investments on performance	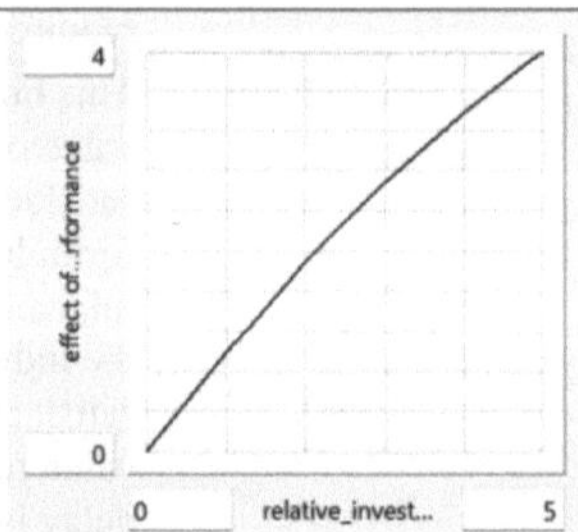	dmnl	The table function effect of investments on performance represents the marginally decreasing positive effect that the allocation of any kind of resources has on the related performance (Frankel, 1962). This function starts from the point (0;0) and reaches the highest value of 4 (that means achieving 100% of the performance required to complete the sustainability transition) when investments are 5 times their normal (5;4). As a result of the calibration and normalisation of the variables, the function passes from the point (1;1) meaning an effect of 1 when investments equal their normal level.

(continued)

Table B.1 (continued)

Variable name	Equation	Units	Documentation
Effect of materiality on expectation	Relative stakeholders' materiality * expectation sensitivity to materiality	dmnl	This variable represents the relationship between stakeholders' materiality and the pressure exerted by stakeholders on the company. When relative materiality is at its maximum, this effect doubles the value of the normal expected performance, making stakeholders' expected performance range between 0 and 1.
Expectation sensitivity to materiality	1	dmnl	This parameter represents the linear effect that a change in stakeholders' materiality level has on stakeholders' expected performance.
Greenwashing performance communication	MAX (0; performance shortfall*greenwashing incentive)	Performance unit	This variable indicates the percentage of shortfall compensated through greenwashing performance communication according to the greenwashing incentive.The max 0 function is used to prevent this variable from going negative when achieved performance is higher than stakeholders' expectations.
Indicated satisfaction	Normal satisfaction * communication effect on satisfaction	Satisfaction unit	This variable represents the goal to which the stock of stakeholders' satisfaction tends to adjust.
Initial satisfaction	0,5	Satisfaction unit	The initial satisfaction value is set to the normal satisfaction value of 0.5: Stakeholders are not satisfied nor dissatisfied by initial business behaviour.It can be changed to simulate a firm entering a new sector with a certain bias due to past behaviours.

(continued)

Table B.1 (continued)

Variable name	Equation	Units	Documentation
Normal dedicated investments	5	Monetary unit	The normal dedicated investment level is set to 5 monetary units according to the outcomes of the model calibration process.
Normal expected performance	0,5	Performance unit	Level of performance that stakeholders expect (0.5) when the materiality level is at its reference value.
Normal performance level	0,25	Performance unit	Level of performance achieved with normal asset level. It is set (through calibration) to 1/4 of the performance corresponding to a complete sustainability transition.
Normal satisfaction	0,5	Satisfaction unit	This normal value represents a satisfaction level of 50%. Stakeholders are indifferent to the business's sustainability behaviour.
Perceived greenwashing	SMTH3(relative greenwashing; perceived greenwashing adjustment time; 0)	dmnl	Greenwashing is hard to detect; for this, a change in the perceived greenwashing is a slow process that requires the information to pass across several sources before being known by stakeholders (de Freitas Netto et al., 2020; Diouf & Boiral, 2017; Jones, 2019) (e.g., suspiciousness, journalists' investigations, information sharing). This situation can be modelled as a third-order information delay through a smooth function SMTH3.

(continued)

Table B.1 (continued)

Variable name	Equation	Units	Documentation
Perceived greenwashing adjustment time	9	Years	The perceived greenwashing adjustment time is set to 9 years, during which the greenwashing consciousness-raising gradually takes place.The 9 years could represent this process (according to the decomposition of the delay structure, Sterman, 2000, pp. 445–448): some distrust could arise after a few years (e.g., 4 years) of communication without solid foundations. This could prompt an investigation by some journalists, who would need to work for additional years (e.g., 3 years) to gather evidence of the behaviour. Similarly, sharing information may not be easy, as the company could continue to defend itself, and offer alternative proofs to the journalists' declarations. Another 2 years might pass before widespread awareness of the false claims is achieved.

(continued)

Table B.1 (continued)

Variable name	Equation	Units	Documentation
Perceived greenwashing multiplier on communication	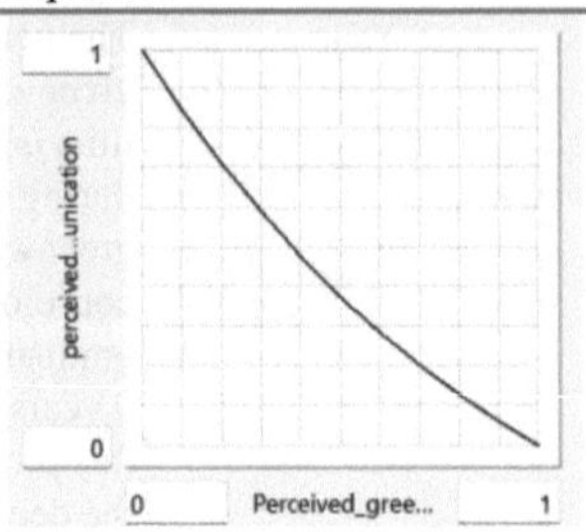	dmnl	This variable is the output of a graphical function used to reproduce the negative consequences of the rise in scepticism. When initial levels of trust are high (i.e., perceived greenwashing = 0), trust can be easily undermined by small negative elements. Indeed, minor violations have a higher marginal impact in the beginning, creating doubt and uncertainty (Lewicki & Bunker, 1996; Slovic, 1993). For this reason, the negative slope of the graphical function is higher at the beginning and keeps decreasing in its absolute value as the perceived greenwashing increases. The multiplier reaches the value of 0 (i.e., the 0% of communication is trusted) when the perceived greenwashing reaches the value of 1 (i.e., the 100%).
Performance shortfall	Stakeholders' expected performance—achieved performance	Performance unit	This variable measures the gap between stakeholders' performance expectations and what the firm achieved (Calabrese et al., 2022). It is not known to stakeholders.
Relative greenwashing	SAFEDIV(greenwashing performance communication; communicated performance; 0)	dmnl	Percentage of total communicated performance deriving from greenwashing practices. When the total communicated performance is 0, also the relative greenwashing goes to 0. It is perceived with a delay by stakeholders.

(continued)

Table B.1 (continued)

Variable name	Equation	Units	Documentation
Relative investment level	Sustainability dedicated assets and competencies / normal dedicated investments	dmnl	Normalisation of sustainability dedicated assets and competencies compared with their normal value.
Satisfaction sensitivity to communication adequacy	2	dmnl	This parameter represents the linear relationship between communication adequacy and satisfaction level. If adequacy is maximum (1), this doubling effect will multiply the normal satisfaction value (0.5).
Stakeholders' expected performance	Effect of materiality on expectation * normal expected performance	Performance unit	This variable represents the level of performance desired by stakeholders. It goes between 0 (no requests) and 1 (maximum efforts requested). The relationship with stakeholders' materiality is assumed to be linear (1:1).
Stakeholders' satisfaction(t)	Stakeholders' satisfaction(t—dt) + (change in satisfaction) * dt	Satisfaction unit	This stock can assume values between 0 and 1. It indicates the stakeholders' level of satisfaction related to the perceived sustainability behaviour of a certain firm.It affects financial achievements because sustainability challenges can represent a competitive advantage for a firm since stakeholders' satisfaction impacts revenues (Morioka et al., 2017).
Time to adjust satisfaction	3	Year	Required time to change stakeholders' satisfaction level. It is set to 3 years and refers to the time needed to change opinion on a firm's overall sustainability behaviour.

(continued)

Table B.1 (continued)

Variable name	Equation	Units	Documentation
"4—Impact materiality sector":			
Average global materiality	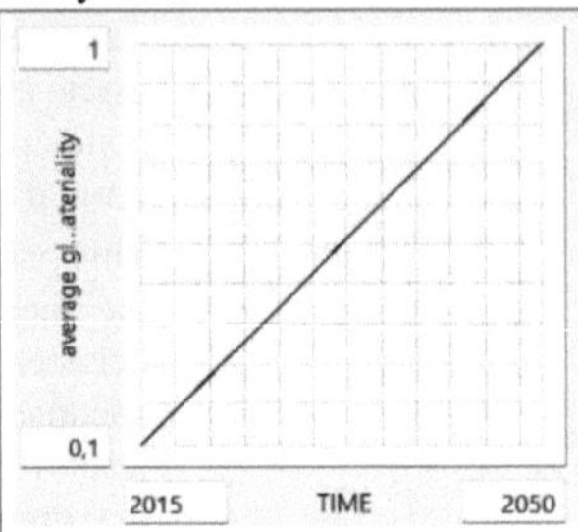	Materiality unit	This variable is based on a linear graphical function that increases from 0 to 1 during the simulation. It represents the increase in the materiality of the sustainability issues according to the SDGs and the European green Deal, with the sustainability transition that needs to be reached in 2050.
Change in stakeholders' materiality	(reference stakeholders' materiality * cumulative effect on stakeholders' materiality—stakeholders' materiality) / time to adjust stakeholders' materiality	Materiality unit/years	This biflow is the yearly rate of change of stakeholders' materiality. It depends on global exogenous aspects and endogenous satisfaction levels. It is structured as an information delay.
Cumulative effect on stakeholders' materiality	Effect of satisfaction on stakeholders' materiality*satisfaction weight + global materiality multiplier on stakeholders' materiality*(1-satisfaction weight)	dmnl	Materiality for stakeholders is affected by two aspects: The first is related to the commitment of the company object of analysis in terms of compliance with stakeholders' expectations; the second is independent of a company's behaviour and concerns the priority of the topic from broader societal and environmental perspectives.

(continued)

Table B.1 (continued)

Variable name	Equation	Units	Documentation
Effect of satisfaction on stakeholders' materiality	Max effect of satisfaction on stakeholders' materiality— (relative satisfaction * sensitivity of materiality to satisfaction)	dmnl	The external materiality of a certain theme is balanced by the business behaviour. If a firm behaves correctly, stakeholders will be satisfied, and their concerns and pressure on the firm to behave better will decrease.This variable is a linear decreasing function with its max equal to the parameter "max effect of satisfaction on stakeholders' materiality" and its angular coefficient equal to the opposite of "sensitivity of materiality to satisfaction".
Global materiality multiplier on stakeholders' materiality	Relative global materiality * sensitivity to global materiality	dmnl	This variable represents the part of impact materiality independent of the company's behaviour.
Max effect of satisfaction on stakeholders' materiality	2	dmnl	The max effect of satisfaction on stakeholders' materiality is set to 2 and represents the maximum multiplier in case of complete dissatisfaction with a company's behaviour (i.e., satisfaction level equal to 0).
Reference satisfaction	0,5	Satisfaction unit	50% reference satisfaction, equal to normal.
Reference stakeholders' materiality	0,5	Materiality unit	50% reference value of stakeholders' materiality
Relative global materiality	Average global materiality/ reference stakeholders' materiality	dmnl	Normalisation of average global materiality according to the reference level of materiality.
Relative satisfaction	Stakeholders' satisfaction/ reference satisfaction	dmnl	Normalisation of stakeholders' satisfaction according to its reference level.

(continued)

Table B.1 (continued)

Variable name	Equation	Units	Documentation
Relative stakeholders' materiality	Stakeholders' materiality / reference stakeholders' materiality	dmnl	The relative stakeholders' materiality is the output of the normalisation (Sterman, 2000, p. 554) of the stock of stakeholders' materiality.
Satisfaction weight	0,5	dmnl	To allow the opportunity to make simulations based on different weights of the two factors, a satisfaction weight variable is added to the mean of the two components and initially set to 0.5
Sensitivity to global materiality	1	dmnl	It is assumed that a change in global average materiality linearly impacts stakeholders' materiality, with a sensitivity of 1.
Sensitivity of materiality to satisfaction	1	dmnl	In line with the approach adopted in the entire model, the negative relationship between satisfaction and materiality is kept linear and simple, assuming that stakeholders' materiality linearly decreases when a positive change in their satisfaction level occurs.
Stakeholders' materiality(t)	Stakeholders' materiality(t—dt) + (change in stakeholders' materiality) * dt	Materiality unit	A theme is material (values between 0 and 1) for a broad range of stakeholders when the company's activities impact the environment and the social dimensions. If a theme is material, stakeholders will increase the pressure on the firm to change its behaviour increasing their expected performance.The initial value of the stock is equal to the average global materiality.

(continued)

Table B.1 (continued)

Variable name	Equation	Units	Documentation
Time to adjust stakeholders' materiality	2	Year	A delay of 2 years represents the resistance of people to change their opinion towards a company or the materiality of an issue.
"5—Financial materiality sector":			
Effect of materiality on market dimension	(average global materiality / reference stakeholders' materiality) * market dimension sensitivity to stakeholders' materiality	dmnl	The global materiality of a theme represents the importance of that sustainability theme. The market size is an indicator of the importance of a certain theme: When it increases, the dimension of the market will have the same behaviour.
Effect of satisfaction on financial achievements	Relative satisfaction*sensitivity of financial achievements to satisfaction	dmnl	The quote of the market dimension caught by the company depends on the level of stakeholders' satisfaction (i.e., relative satisfaction) multiplied by the sensitivity of financial achievements to satisfaction.
Maximum market dimension	4	dmnl	The maximum market dimension, set equal to 4 for calibration reasons, represents the maximum achievable financial achievements when average global materiality is equal to 1.
Sustainability-related market dimension	Effect of materiality on market dimension * maximum market dimension	dmnl	In this model, the size of the market is measured in relative terms, compared with the actual dimension of the firm. This variable represents, for example, the potential financial benefits achievable, dependent on the average global materiality of an issue.

(continued)

Table B.1 (continued)

Variable name	Equation	Units	Documentation
Financial achievements (or financial materiality)	Sustainability-related market dimension * effect of satisfaction on financial achievements	dmnl	This variable shows the yearly achieved positive financial repercussions of the sustainability performance. It should be seen as a percentage of its initial business-as-usual performance.It also represents the financial materiality of an issue, impacting the internal materiality sector through the materiality assessment process.
Sensitivity of financial achievements to satisfaction	0,5	dmnl	Being the relative level of satisfaction = 0.5, this sensitivity parameter keeps the level of financial achievements = 0.5 when satisfaction is at its normal.
Market dimension sensitivity to stakeholders' materiality	0,5	dmnl	This sensitivity set to 0.5 makes the effect range between 0 and 1.

Source: Author's elaboration

Table B.2 Simulation settings

Run Specs			
Start Time	2015	Pause Interval	0
Stop time	2050	Integration method	Euler
DT	1/8	Keep all variable results	True
Fractional DT	True	Run by	Run
Save interval	0,125	Calculate loop dominance information	True
Sim duration	1,5	Exhaustive search threshold	1000
Time units	Years		

Source: Author's elaboration

Glossary

CDP	Carbon Disclosure Project
CLD	Causal Loop Diagram
COP	Conference Of Parties
CSR	Corporate Social Responsibility
CSRD	Corporate Sustainability Reporting Directive
dmnl	Dimensionless
DPM	Dynamic Performance Management
EFRAG	European Financial Advisory Group
ESG	Environmental Social Governance
ESRS	European Sustainability Reporting Standards
EU	European Union
GHG	Greenhouse Gasses
GRI	Global Reporting Initiative
IAS	International Accounting Standards
IASB	International Accounting Standards Board
IFRS	International Financial Reporting Standards
IIRC	International Integrated Reporting Council
ISO	International Standard Organisation
ISSB	International Sustainability Standards Board
KPI	Key Performance Indicator
MDGs	Millennium Development Goals
MIT	Massachusetts Institute of Technology
NFRD	Non-Financial Reporting Directive
NGOs	Non-Governmental Organisations
OECD	Organization for Economic Co-operation and Development
OIBR	Organismo Italiano di Business Reporting
ROI	Return On Investments
SAI	Social Accountability International
SASB	Sustainability Accounting Standard Board

R. Santoni, *The Dynamics of Materiality*, System Dynamics for Performance Management & Governance 8, https://doi.org/10.1007/978-3-032-19280-6

SDGs	Sustainable Development Goals
SEC	Securities and Exchange Commission
SFD	Stock and Flow Diagram
SMEs	Small and Medium Enterprises
SROI	Social/Sustainability Return On Investments
TCFD	Task force on Climate-related Financial Disclosure
UN	United Nations
UNCTAD	UN Conference on Trade and Development
UNEP	UN Environment Programme
UNGC	UN Global Compact
VRF	Value Reporting Foundation
WCED	World Commission on Environment and Development

MIX
Papier aus verantwortungsvollen Quellen
Paper from responsible sources
FSC® C105338

If you have any concerns about our products,
you can contact us on
ProductSafety@springernature.com

In case Publisher is established outside the EU,
the EU authorized representative is:
Springer Nature Customer Service Center GmbH
Europaplatz 3, 69115 Heidelberg, Germany

Printed by Libri Plureos GmbH
in Hamburg, Germany